Pitman Research Notes in Mathematics Series

Submission of proposals for consideration

Suggestions for publication, in the form of outlines and representative samples, are invited by the Editorial Board for assessment. Intending authors should approach one of the main editors or another member of the Editorial Board, citing the relevant AMS subject classifications. Alternatively, outlines may be sent directly to the publisher's offices. Refereeing is by members of the board and other mathematical authorities in the topic concerned, throughout the world.

Preparation of accepted manuscripts

On acceptance of a proposal, the publisher will supply full instructions for the preparation of manuscripts in a form suitable for direct photo-lithographic reproduction. Specially printed grid sheets can be provided and a contribution is offered by the publisher towards the cost of typing. Word processor output, subject to the publisher's approval, is also acceptable.

Illustrations should be prepared by the authors, ready for direct reproduction without further improvement. The use of hand-drawn symbols should be avoided wherever possible, in order to maintain maximum clarity of the text.

The publisher will be pleased to give any guidance necessary during the preparation of a typescript, and will be happy to answer any queries.

Important note

In order to avoid later retyping, intending authors are strongly urged not to begin final preparation of a typescript before receiving the publisher's guidelines. In this way it is hoped to preserve the uniform appearance of the series.

Longman Scientific & Technical
Longman House
Burnt Mill
Harlow, Essex, CM20 2JE
UK
(Telephone (0279) 426721)

Titles in this series. A full list is available from the publisher on request.

51 Subnormal operators
J B Conway

52 Wave propagation in viscoelastic media
F Mainardi

53 Nonlinear partial differential equations and their applications: Collège de France Seminar. Volume I
H Brezis and J L Lions

54 Geometry of Coxeter groups
H Hiller

55 Cusps of Gauss mappings
T Banchoff, T Gaffney and C McCrory

56 An approach to algebraic K-theory
A J Berrick

57 Convex analysis and optimization
J-P Aubin and R B Vintner

58 Convex analysis with applications in the differentiation of convex functions
J R Giles

59 Weak and variational methods for moving boundary problems
C M Elliott and J R Ockendon

60 Nonlinear partial differential equations and their applications: Collège de France Seminar. Volume II
H Brezis and J L Lions

61 Singular Systems of differential equations II
S L Campbell

62 Rates of convergence in the central limit theorem
Peter Hall

63 Solution of differential equations by means of one-parameter groups
J M Hill

64 Hankel operators on Hilbert Space
S C Power

65 Schrödinger-type operators with continuous spectra
M S P Eastham and H Kalf

66 Recent applications of generalized inverses
S L Campbell

67 Riesz and Fredholm theory in Banach algebra
B A Barnes, G J Murphy, M R F Smyth and T T West

68 Evolution equations and their applications
K Kappel and W Schappacher

69 Generalized solutions of Hamilton–Jacobi equations
P L Lions

70 Nonlinear partial differential equations and their applications: Collège de France Seminar. Volume III
H Brezis and J L Lions

71 Spectral theory and wave operators for the Schrödinger equation
A M Berthier

72 Approximation of Hilbert space operators I
D A Herrero

73 Vector valued Nevanlinna theory
H J W Ziegler

74 Instability, nonexistence and weighted energy methods in fluid dynamics and related theories
B Straughan

75 Local bifurcation and symmetry
A Vanderbauwhede

76 Clifford analysis
F Brackx, R Delanghe and F Sommen

77 Nonlinear equivalence, reduction of PDEs to ODEs and fast convergent numerical methods
E E Rosinger

78 Free boundary problems, theory and applications. Volume I
A Fasano and M Primicerio

79 Free boundary problems, theory and applications. Volume II
A Fasano and M Primicerio

80 Symplectic geometry
A Crumeyrolle and J Grifone

81 An algorithmic analysis of a communication model with retransmission of flawed messages
D M Lucantoni

82 Geometric games and their applications
W H Ruckle

83 Additive groups of rings
S Feigelstock

84 Nonlinear partial differential equations and their applications: Collège de France Seminar. Volume IV
H Brezis and J L Lions

85 Multiplicative functionals on topological algebras
T Husain

86 Hamilton–Jacobi equations in Hilbert spaces
V Barbu and G Da Prato

87 Harmonic maps with symmetry, harmonic morphisms and deformations of metric
P Baird

88 Similarity solutions of nonlinear partial differential equations
L Dresner

89 Contributions to nonlinear partial differential equations
C Bardos, A Damlamian, J I Díaz and J Hernández

90 Banach and Hilbert spaces of vector-valued functions
J Burbea and P Masani

91 Control and observation of neutral systems
D Salamon

92 Banach bundles, Banach modules and automorphisms of C^*-algebras
M J Dupré and R M Gillette

93 Nonlinear partial differential equations and their applications: Collège de France Seminar. Volume V
H Brezis and J L Lions

94 Computer algebra in applied mathematics: an introduction to MACSYMA
R H Rand

95 Advances in nonlinear waves. Volume I
L Debnath

96 FC-groups
M J Tomkinson

97 Topics in relaxation and ellipsoidal methods
M Akgül

98 Analogue of the group algebra for topological semigroups
H Dzinotyiweyi

99 Stochastic functional differential equations
S E A Mohammed

100 Optimal control of variational inequalities
V Barbu
101 Partial differential equations and dynamical systems
W E Fitzgibbon III
102 Approximation of Hilbert space operators Volume II
C Apostol, L A Fialkow, D A Herrero and D Voiculescu
103 Nondiscrete induction and iterative processes
V Ptak and F-A Potra
104 Analytic functions – growth aspects
O P Juneja and G P Kapoor
105 Theory of Tikhonov regularization for Fredholm equations of the first kind
C W Groetsch
106 Nonlinear partial differential equations and free boundaries. Volume I
J I Díaz
107 Tight and taut immersions of manifolds
T E Cecil and P J Ryan
108 A layering method for viscous, incompressible L_p flows occupying R^n
A Douglis and E B Fabes
109 Nonlinear partial differential equations and their applications: Collège de France Seminar. Volume VI
H Brezis and J L Lions
110 Finite generalized quadrangles
S E Payne and J A Thas
111 Advances in nonlinear waves. Volume II
L Debnath
112 Topics in several complex variables
E Ramírez de Arellano and D Sundararaman
113 Differential equations, flow invariance and applications
N H Pavel
114 Geometrical combinatorics
F C Holroyd and R J Wilson
115 Generators of strongly continuous semigroups
J A van Casteren
116 Growth of algebras and Gelfand–Kirillov dimension
G R Krause and T H Lenagan
117 Theory of bases and cones
P K Kamthan and M Gupta
118 Linear groups and permutations
A R Camina and E A Whelan
119 General Wiener–Hopf factorization methods
F-O Speck
120 Free boundary problems: applications and theory. Volume III
A Bossavit, A Damlamian and M Fremond
121 Free boundary problems: applications and theory. Volume IV
A Bossavit, A Damlamian and M Fremond
122 Nonlinear partial differential equations and their applications: Collège de France Seminar. Volume VII
H Brezis and J L Lions
123 Geometric methods in operator algebras
H Araki and E G Effros
124 Infinite dimensional analysis–stochastic processes
S Albeverio
125 Ennio de Giorgi Colloquium
P Krée
126 Almost-periodic functions in abstract spaces
S Zaidman
127 Nonlinear variational problems
A Marino, L Modica, S Spagnolo and M Degliovanni
128 Second-order systems of partial differential equations in the plane
L K Hua, W Lin and C-Q Wu
129 Asymptotics of high-order ordinary differential equations
R B Paris and A D Wood
130 Stochastic differential equations
R Wu
131 Differential geometry
L A Cordero
132 Nonlinear differential equations
J K Hale and P Martinez-Amores
133 Approximation theory and applications
S P Singh
134 Near-rings and their links with groups
J D P Meldrum
135 Estimating eigenvalues with *a posteriori/a priori* inequalities
J R Kuttler and V G Sigillito
136 Regular semigroups as extensions
F J Pastijn and M Petrich
137 Representations of rank one Lie groups
D H Collingwood
138 Fractional calculus
G F Roach and A C McBride
139 Hamilton's principle in continuum mechanics
A Bedford
140 Numerical analysis
D F Griffiths and G A Watson
141 Semigroups, theory and applications. Volume I
H Brezis, M G Crandall and F Kappel
142 Distribution theorems of L-functions
D Joyner
143 Recent developments in structured continua
D De Kee and P Kaloni
144 Functional analysis and two-point differential operators
J Locker
145 Numerical methods for partial differential equations
S I Hariharan and T H Moulden
146 Completely bounded maps and dilations
V I Paulsen
147 Harmonic analysis on the Heisenberg nilpotent Lie group
W Schempp
148 Contributions to modern calculus of variations
L Cesari
149 Nonlinear parabolic equations: qualitative properties of solutions
L Boccardo and A Tesei
150 From local times to global geometry, control and physics
K D Elworthy

151 A stochastic maximum principle for optimal control of diffusions
U G Haussmann
152 Semigroups, theory and applications. Volume II
H Brezis, M G Crandall and F Kappel
153 A general theory of integration in function spaces
P Muldowney
154 Oakland Conference on partial differential equations and applied mathematics
L R Bragg and J W Dettman
155 Contributions to nonlinear partial differential equations. Volume II
J I Díaz and P L Lions
156 Semigroups of linear operators: an introduction
A C McBride
157 Ordinary and partial differential equations
B D Sleeman and R J Jarvis
158 Hyperbolic equations
F Colombini and M K V Murthy
159 Linear topologies on a ring: an overview
J S Golan
160 Dynamical systems and bifurcation theory
M I Camacho, M J Pacifico and F Takens
161 Branched coverings and algebraic functions
M Namba
162 Perturbation bounds for matrix eigenvalues
R Bhatia
163 Defect minimization in operator equations: theory and applications
R Reemtsen
164 Multidimensional Brownian excursions and potential theory
K Burdzy
165 Viscosity solutions and optimal control
R J Elliott
166 Nonlinear partial differential equations and their applications: Collège de France Seminar. Volume VIII
H Brezis and J L Lions
167 Theory and applications of inverse problems
H Haario
168 Energy stability and convection
G P Galdi and B Straughan
169 Additive groups of rings. Volume II
S Feigelstock
170 Numerical analysis 1987
D F Griffiths and G A Watson
171 Surveys of some recent results in operator theory. Volume I
J B Conway and B B Morrel
172 Amenable Banach algebras
J-P Pier
173 Pseudo-orbits of contact forms
A Bahri
174 Poisson algebras and Poisson manifolds
K H Bhaskara and K Viswanath
175 Maximum principles and eigenvalue problems in partial differential equations
P W Schaefer
176 Mathematical analysis of nonlinear, dynamic processes
K U Grusa
177 Cordes' two-parameter spectral representation theory
D F McGhee and R H Picard
178 Equivariant K-theory for proper actions
N C Phillips
179 Elliptic operators, topology and asymptotic methods
J Roe
180 Nonlinear evolution equations
J K Engelbrecht, V E Fridman and E N Pelinovski
181 Nonlinear partial differential equations and their applications: Collège de France Seminar. Volume IX
H Brezis and J L Lions
182 Critical points at infinity in some variational problems
A Bahri
183 Recent developments in hyperbolic equations
L Cattabriga, F Colombini, M K V Murthy and S Spagnolo
184 Optimization and identification of systems governed by evolution equations on Banach space
N U Ahmed
185 Free boundary problems: theory and applications. Volume I
K H Hoffmann and J Sprekels
186 Free boundary problems: theory and applications. Volume II
K H Hoffmann and J Sprekels
187 An introduction to intersection homology theory
F Kirwan
188 Derivatives, nuclei and dimensions on the frame of torsion theories
J S Golan and H Simmons
189 Theory of reproducing kernels and its applications
S Saitoh
190 Volterra integrodifferential equations in Banach spaces and applications
G Da Prato and M Iannelli
191 Nest algebras
K R Davidson
192 Surveys of some recent results in operator theory. Volume II
J B Conway and B B Morrel
193 Nonlinear variational problems. Volume II
A Marino and M K V Murthy
194 Stochastic processes with multidimensional parameter
M E Dozzi
195 Prestressed bodies
D Iesan
196 Hilbert space approach to some classical transforms
R H Picard
197 Stochastic calculus in application
J R Norris
198 Radical theory
B J Gardner
199 The C^*-algebras of a class of solvable Lie groups
X Wang
200 Stochastic analysis, path integration and dynamics
K D Elworthy and J C Zambrini

201 Riemannian geometry and holonomy groups
S Salamon
202 Strong asymptotics for extremal errors and polynomials associated with Erdös type weights
D S Lubinsky
203 Optimal control of diffusion processes
V S Borkar
204 Rings, modules and radicals
B J Gardner
205 Two-parameter eigenvalue problems in ordinary differential equations
M Faierman
206 Distributions and analytic functions
R D Carmichael and D Mitrovic
207 Semicontinuity, relaxation and integral representation in the calculus of variations
G Buttazzo
208 Recent advances in nonlinear elliptic and parabolic problems
P Bénilan, M Chipot, L Evans and M Pierre
209 Model completions, ring representations and the topology of the Pierce sheaf
A Carson
210 Retarded dynamical systems
G Stepan
211 Function spaces, differential operators and nonlinear analysis
L Paivarinta
212 Analytic function theory of one complex variable
C C Yang, Y Komatu and K Niino
213 Elements of stability of visco-elastic fluids
J Dunwoody
214 Jordan decomposition of generalized vector measures
K D Schmidt
215 A mathematical analysis of bending of plates with transverse shear deformation
C Constanda
216 Ordinary and partial differential equations. Volume II
B D Sleeman and R J Jarvis
217 Hilbert modules over function algebras
R G Douglas and V I Paulsen
218 Graph colourings
R Wilson and R Nelson
219 Hardy-type inequalities
A Kufner and B Opic
220 Nonlinear partial differential equations and their applications: Collège de France Seminar. Volume X
H Brezis and J L Lions
221 Workshop on dynamical systems
E Shiels and Z Coelho
222 Geometry and analysis in nonlinear dynamics
H W Broer and F Takens
223 Fluid dynamical aspects of combustion theory
M Onofri and A Tesei
224 Approximation of Hilbert space operators. Volume I. 2nd edition
D Herrero
225 Operator theory: proceedings of the 1988 GPOTS–Wabash conference
J B Conway and B B Morrel
226 Local cohomology and localization
J L Bueso Montero, B Torrecillas Jover and A Verschoren
227 Nonlinear waves and dissipative effects
D Fusco and A Jeffrey
228 Numerical analysis 1989
D F Griffiths and G A Watson
229 Recent developments in structured continua. Volume II
D De Kee and P Kaloni
230 Boolean methods in interpolation and approximation
F J Delvos and W Schempp
231 Further advances in twistor theory. Volume I
L J Mason and L P Hughston
232 Further advances in twistor theory. Volume II
L J Mason and L P Hughston
233 Geometry in the neighborhood of invariant manifolds of maps and flows and linearization
U Kirchgraber and K Palmer
234 Quantales and their applications
K I Rosenthal
235 Integral equations and inverse problems
V Petkov and R Lazarov
236 Pseudo-differential operators
S R Simanca
237 A functional analytic approach to statistical experiments
I M Bomze
238 Quantum mechanics, algebras and distributions
D Dubin and M Hennings
239 Hamilton flows and evolution semigroups
J Gzyl
240 Topics in controlled Markov chains
V S Borkar
241 Invariant manifold theory for hydrodynamic transition
S Sritharan
242 Lectures on the spectrum of $L^2(\Gamma\backslash G)$
F L Williams
243 Progress in variational methods in Hamiltonian systems and elliptic equations
M Girardi, M Matzeu and F Pacella
244 Optimization and nonlinear analysis
A Ioffe, M Marcus and S Reich
245 Inverse problems and imaging
G F Roach
246 Semigroup theory with applications to systems and control
N U Ahmed
247 Periodic-parabolic boundary value problems and positivity
P Hess
248 Distributions and pseudo-differential operators
S Zaidman
249 Progress in partial differential equations: the Metz surveys
M Chipot and J Saint Jean Paulin
250 Differential equations and control theory
V Barbu

251 Stability of stochastic differential equations with respect to semimartingales
X Mao
252 Fixed point theory and applications
J Baillon and M Théra
253 Nonlinear hyperbolic equations and field theory
M K V Murthy and S Spagnolo
254 Ordinary and partial differential equations. Volume III
B D Sleeman and R J Jarvis
255 Harmonic maps into homogeneous spaces
M Black
256 Boundary value and initial value problems in complex analysis: studies in complex analysis and its applications to PDEs 1
R Kühnau and W Tutschke
257 Geometric function theory and applications of complex analysis in mechanics: studies in complex analysis and its applications to PDEs 2
R Kühnau and W Tutschke
258 The development of statistics: recent contributions from China
X R Chen, K T Fang and C C Yang
259 Multiplication of distributions and applications to partial differential equations
M Oberguggenberger
260 Numerical analysis 1991
D F Griffiths and G A Watson
261 Schur's algorithm and several applications
M Bakonyi and T Constantinescu
262 Partial differential equations with complex analysis
H Begehr and A Jeffrey
263 Partial differential equations with real analysis
H Begehr and A Jeffrey
264 Solvability and bifurcations of nonlinear equations
P Drábek
265 Orientational averaging in mechanics of solids
A Lagzdins, V Tamuzs, G Teters and A Kregers
266 Progress in partial differential equations: elliptic and parabolic problems
C Bandle, J Bemelmans, M Chipot, M Grüter and J Saint Jean Paulin
267 Progress in partial differential equations: calculus of variations, applications
C Bandle, J Bemelmans, M Chipot, M Grüter and J Saint Jean Paulin
268 Stochastic partial differential equations and applications
G Da Prato and L Tubaro
269 Partial differential equations and related subjects
M Miranda
270 Operator algebras and topology
W B Arveson, A S Mishchenko, M Putinar, M A Rieffel and S Stratila
271 Operator algebras and operator theory
W B Arveson, A S Mishchenko, M Putinar, M A Rieffel and S Stratila
272 Ordinary and delay differential equations
J Wiener and J K Hale
273 Partial differential equations
J Wiener and J K Hale
274 Mathematical topics in fluid mechanics
J F Rodrigues and A Sequeira
275 Green functions for second order parabolic integro-differential problems
M G Garroni and J F Menaldi
276 Riemann waves and their applications
M W Kalinowski
277 Banach C(K)-modules and operators preserving disjointness
Y A Abramovich, E L Arenson and A K Kitover
278 Limit algebras: an introduction to subalgebras of C*-algebras
S C Power
279 Abstract evolution equations, periodic problems and applications
D Daners and P Koch Medina
280 Emerging applications in free boundary problems
J Chadam and H Rasmussen
281 Free boundary problems involving solids
J Chadam and H Rasmussen
282 Free boundary problems in fluid flow with applications
J Chadam and H Rasmussen
283 Asymptotic problems in probability theory: stochastic models and diffusions on fractals
K D Elworthy and N Ikeda
284 Asymptotic problems in probability theory: Wiener functionals and asymptotics
K D Elworthy and N Ikeda
285 Dynamical systems
R Bamon, R Labarca, J Lewowicz and J Palis
286 Models of hysteresis
A Visintin
287 Moments in probability and approximation theory
G A Anastassiou
288 Mathematical aspects of penetrative convection
B Straughan
289 Ordinary and partial differential equations. Volume IV
B D Sleeman and R J Jarvis
290 K-theory for real C^*-algebras
H Schröder
291 Recent developments in theoretical fluid mechanics
G P Galdi and J Necas
292 Propagation of a curved shock and nonlinear ray theory
P Prasad
293 Non-classical elastic solids
M Ciarletta and D Ieşan
294 Multigrid methods
J Bramble
295 Entropy and partial differential equations
W A Day
296 Progress in partial differential equations: the Metz surveys 2
M Chipot
297 Nonstandard methods in the calculus of variations
C Tuckey
298 Barrelledness, Baire-like- and (LF)-spaces
M Kunzinger

Michael Kunzinger

University of Vienna, Austria

Barrelledness, Baire-like- and (LF)-spaces

Copublished in the United States with
John Wiley & Sons, Inc., New York

Longman Scientific & Technical
Longman Group UK Limited
Longman House, Burnt Mill, Harlow
Essex CM20 2JE, England
and Associated companies throughout the world.

Copublished in the United States with
John Wiley & Sons Inc., 605 Third Avenue, New York, NY 10158

First published 1993

AMS Subject Classifications: 46A08, 46A13

ISSN 0269-3674

ISBN 0 582 23745 9

British Library Cataloguing in Publication Data

A catalogue record for this book is
available from the British Library

Library of Congress Cataloging-in-Publication Data

Kunzinger, M. (Michael)
Barrelledness, Baire-like- and (LF)-spaces / M. Kunzinger.
p. cm. -- (Pitman research notes in mathematics series, ISSN 0269-3674 ;)
Includes bibliographical references and index.
1. Barrelled spaces. 2. Baire spaces. 3. Locally convex spaces.
I. Title. II. Series.
QA322.K86 1993
515'.73--dc20 93-35711
CIP

Printed and bound in Great Britain
by Biddles Ltd, Guildford and King's Lynn

Contents

Preface

The purpose of the present book is to present a modern approach to a development on the field of locally convex spaces that took place over the last twenty years. The roots of this subject lie in a number of weakened barrelledness concepts that are introduced in the first chapter, where we derive a number of methods and results that will turn out to be of great value in subsequent chapters.

During the research that led to the writing of this book we repeatedly came across (infinite-dimensional) locally convex spaces carrying the strongest locally convex topology. For quite some time such spaces have been considered to be only of theoretical interest (e.g. for constructing counterexamples or to display certain pathologies of locally convex spaces), but not to have any importance for analysis-related problems. However, they proved to play a decisive role both in the study of weakenings of the Baire property in chapters 5 and 6 and on the field of generalized inductive limits as presented in chapter 7. These facts led us to dedicate an entire chapter to the spaces in question.

Another section on varieties of locally convex spaces was included mainly for aesthetic reasons (that is, because we think it's a beautiful theory). The concept of varieties is intended to be a unifying element in this work and to present other lines of research that grow out of the same origins as the subjects we are essentially concerned with.

A major part of the book deals with several classes of spaces that are 'more than barrelled' but 'less than Baire'. In a chapter on linear Baire spaces at the beginning of these considerations we present some characterizations of the Baire property that are usually not included in textbooks on locally convex spaces.

Apart from the importance of 'almost Baire' spaces for questions of classifying inductive limits of locally convex spaces we basically cover this field for two reasons:

First, these spaces show a number of optimal permanence properties that are not known to (or do not) hold for locally convex Baire spaces. We hope that the detailed analysis of these characteristics will help to clarify the situation for locally convex Baire spaces themselves.

In addition to this, 'almost Baire spaces' can be employed to derive several generalizations of well-known open-mapping and closed-graph theorems.

Concerning inheritance properties, the following observation is worth pointing out: Practically every class of spaces examined here is stable under the operation of taking countable-codimensional subspaces.

The final section of the present book is dedicated to the study of inductive limits of locally convex spaces, particularly (LF)- and (LB)-spaces. Apart from a classification of these spaces by means of the weakened Baire properties introduced in the chapters before, it presents several quite recent results on metrizable and normable (LF)-spaces. For instance, we prove here that many of the classical Banach spaces contain dense (LF)-subspaces. These questions are closely related to the (still unsolved) separable quotient problem in Banach spaces as will also be demonstrated.

Although certain aspects of the subject of this book are also treated in the excellent monographs [56] by M.VALDIVIA and [5] by J.BONET and P.PEREZ CARRERAS (to which we refer the reader for further study), we believe that we present here for the first time a self-contained introduction to this theory (especially concerning a modern exposition of the theory of (LF)-spaces and the interrelation of generalized inductive limits, 'almost barrelled' and 'almost Baire' spaces). The book is self-contained insofar as all results (with few exceptions such as certain examples) are proved on a level that connects this treatise with the standard textbooks on the subject (particularly J.HORVATH ([23]) and H.H.SCHÄFER ([50]).

It is most important to me to thank professor MICHAEL GROSSER for the immense amount of time and effort he spent on this project. Without his expert knowledge, creativity and endurance, the book in its present form would not have been possible. All those parts of this work containing original or new results are decisively influenced by his ideas. So whenever the term 'we' is used within the text it should be understood as an abbreviation for 'M.GROSSER and me'.

I am also indebted to professor M.OBERGUGGENBERGER from the university of Innsbruck for permanent help and encouragement. Finally, I would like to thank professor S.A.SAXON from the university of Florida - whose works have inspired large parts of this book - for several helpful comments that contributed to the final version of the manuscript.

Vienna, September 1993 Michael Kunzinger

Notations

Generally following the terminology of J.HORVATH ([23]) and H.H.SCHÄFER ([50]), we only specify some particularly important definitions:
By an *lcs* we mean a locally convex *Hausdorff* space.
A *tvs* is a (not necessarily Hausdorff) topological vector space. When we speak of a vector space with a locally convex topology, this space also doesn't have to be separated.
We denote by E^* the algebraical dual and by E' the topological dual of any given tvs E.
The word 'countable' comprises both the finite and the coutably infinite case.
All tvs's treated here are real or complex, i.e. we always have $\mathbf{K} = \mathbf{R}$ or $\mathbf{K} = \mathbf{C}$.

1 Weakenings of the Concept of Barrelledness

In this chapter we intend to give an introduction to the structural theory of several classes of 'almost barrelled' spaces. The definitions and techniques developed in this section provide the foundation on which both the theory of weakened Baire properties and the study of generalized inductive limits of lcs's rest.
All the concepts that are being dealt with will be characterized in terms of duality as well as through comparison of suitable locally convex topologies. Apart from this, we are interested in the interdependency of the various generalizations of barrelledness. In particular, we will examine additional conditions on the underlying lcs (such as metrizability or separability) under which some of the barrelledness concepts in question coincide. On the other hand, a list of distinguishing examples in 1.9 will demonstrate that, in general, none of these notions coincide.
Another important point is the study of permanence properties: It will turn out (here and in chapters 5 and 6) that 'almost barrelled' and 'almost Baire' spaces show remarkably similar stability properties, especially concerning the formation of countable-codimensional subspaces.

1.1 Definitions

First of all, let us remind the reader of some basic facts on barrelled spaces:
A *barrel* in a tvs E is a subset which is absolutely convex, closed and absorbing. By polarity, barrels are just the polars in E of bounded subsets of $(E', \sigma(E', E))$. An lcs E is called *barrelled* if each barrel in E is a neighborhood of the origin. E is barrelled if and only if the following families of subsets of E' are identical:

(i) the equicontinuous sets

(ii) the relatively $\sigma(E', E)$-compact sets

(iii) the $\beta(E', E)$-bounded sets

(iv) the $\sigma(E', E)$-bounded sets

(cf.[23],p.212 or [50], IV.5.2).

Separated quotients, locally convex direct sums and Hausdorff inductive limits of barrelled spaces are barrelled(see [50],II.7.2.). Every product of barrelled spaces is barrelled(see [50],IV.4.3,Cor.3).

Since finite- (and countable-) codimensional subspaces will play an important role throughout this book, we explicitely state a proof of the following well known fact:

1.1.1 Proposition *A finite-codimensional subspace of a barrelled space is barrelled.*

Proof. By induction we may assume without loss of generality that F has codimension 1 in E, so that there exists an $x \in E$ with $E = F + sp(x)$. If B is a barrel in F, we define

$$A := \begin{cases} B + \{\lambda x \mid | \lambda | \leq 1\} & \text{if } B = \overline{B} \\ \overline{B} & \text{if } B \neq \overline{B} \end{cases}$$

A is a barrel in E: This is clear if $B = \overline{B}$.In the second case we only have to show that x is absorbed by $A = \overline{B}$.Choose $z \in \overline{B} \setminus B \subseteq E \setminus F$ (B is closed in F!). Then $z = y + \lambda x$ for some $\lambda \neq 0$, $y \in F$.Now if $0 < \rho \leq 1$ is such that $\rho y \in B$, we have:

$$\frac{1}{2}\rho\lambda x = \frac{1}{2}\rho z - \frac{1}{2}\rho y \in \frac{1}{2}\overline{B} - \frac{1}{2}\overline{B} \subseteq \overline{B}.$$

Since E is barrelled, A is a neighborhood of 0 in E. The fact that $B = A \cap F$ completes the proof. ■

The proof of 1.1.1, together with finite induction, immediately yields:

1.1.2 Corollary *Let E be an lcs, F a finite-codimensional subspace of E and B a barrel in F. Then there exists a barrel B' in E such that $B = B' \cap F$.* ■

There are several generalizations of the notion 'barrelled', some of which we present in

1.1.3 Definition *Let E be an lcs*

An infrabarrel *in E is a barrel which is the intersection of a sequence of closed and absolutely convex neighborhoods of 0.*

An ω-barrel *is the polar in E of a bounded sequence in $(E', \sigma(E', E))$ or, equivalently, a barrel which is the polar of a sequence in $(E', \sigma(E', E))$. E is called* infrabarrelled *(resp.*ω-barrelled*) if every infrabarrel (resp. ω-barrel) in E is a neighborhood of 0.*

E is evaluable *(resp.*infraevaluable*,resp.*ω-evaluable*) if every bornivorous barrel*

(resp.infrabarrel,resp.ω-barrel) in E is a neighborhood of 0.

1.1.4 Remark If $B = \{f_n \mid n \in \mathbf{N}\}^\circ$ is an ω-barrel,then $B = \bigcap_{n=1}^{\infty}\{f_n\}^\circ$ is also an infrabarrel and, in particular, a barrel.
Thus we have the following implications:

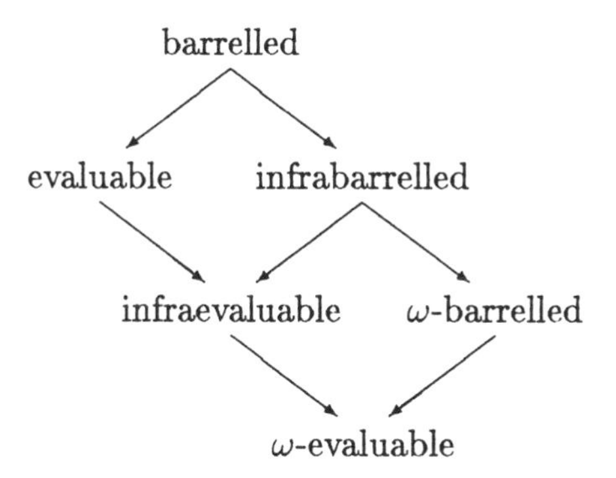

1.2 Dual Characterization

Each of the five properties in 1.1.3 (as well as the term 'barrelled' itself) is determined by the inclusion of certain classes of subsets of E in others. Each of these classes, in turn, consists of sets of a certain kind , described by one or two items from the '(x)-list' below, where $x \in \{\alpha, \beta, \gamma, \delta, \epsilon\}$.
By the bipolar theorem ([50], IV.1.5) we obtain conditions equivalent to the six properties in question if we pass to the polars of the respective subsets of E and state inclusion relations for the respective classes of subsets of E', described by the appropriate items from the '(x')-list', also given below.
In a second step, we again obtain equivalent conditions by dropping the restrictions on the (x')-sets to be $\sigma(E', E)$-closed and absolutely convex. This passage to the resulting properties (x'') is based on the fact that each (x'')-set is contained in an (x')-set of the respective kind and that subsets of equicontinuous sets are equicontinuous.
In the following, we present the (x)- resp. (x')- resp. (x'')-list and give arguments for the propositions (P_x): 'A closed and absolutely convex set W is an (x)-set iff W° is an (x')-set.' and (Q_x): 'Each (x'')-set is contained in some (x')-set' ($x \in \{\alpha, \beta, \gamma, \delta, \epsilon\}$).
We abbreviate the term '$(\sigma(E', E)$-) closed and absolutely convex' by 'clac' (resp.'σ-clac').

(α) clac neighborhood of 0;

(β) bornivorous barrel;

(γ) barrel;

(δ) $\bigcap_{n=1}^{\infty} V_n$, V_n a clac neighborhood of 0 for each n;

(ϵ) $\{f_n \mid n \in \mathbf{N}\}^\circ$, $f_n \in E'$;

(α') σ-clac equicontinuous set;

(β') σ-clac $\beta(E', E)$-bounded set;

(γ') σ-clac $\sigma(E', E)$-bounded set;

(δ') $\sigma(E', E)$-closure of $\Gamma_{n=1}^{\infty} H_n$, H_n equicontinuous;

(ϵ') $\sigma(E', E)$-closure of $\Gamma_{n=1}^{\infty}\{f_n\}$, $f_n \in E'$;

(α'') equicontinuous set;

(β'') $\beta(E', E)$-bounded set;

(γ'') $\sigma(E', E)$-bounded set;

(δ'') $\bigcup_{n=1}^{\infty} H_n$, H_n equicontinuous;

(ϵ'') $\{f_n \mid n \in \mathbf{N}\}$, $f_n \in E'$;

For (P_α), see [50], p.125, remark 5; for (P_β), see [50], p.142; for (P_γ), see [50], IV.1.6. Concerning (P_δ), we supply a proof:
If $V = \bigcap_{n=1}^{\infty} V_n$ is as in (δ), then IV.1.5, Cor.2 in [50] gives $V^\circ = \overline{\Gamma}_{n=1}^{\infty} V_n^\circ$. Conversely, if $D = \overline{\Gamma}_{n=1}^{\infty} H_n$ is a (δ')-set, then $H_n^{\circ\circ} = \overline{\Gamma}_{n=1}^{\infty} H_n$ is σ-clac and equicontinuous for each n. Observing $H_n \subseteq H_n^{\circ\circ} = \overline{\Gamma}_{n=1}^{\infty} H_n \subseteq D$, [50], IV.1.5 Cor.2, again, implies

$$D = \overline{\Gamma}_{n=1}^{\infty} H_n^{\circ\circ} = \left(\bigcap_{n=1}^{\infty} H_n^\circ\right)^\circ = \left(\bigcap_{n=1}^{\infty} V_n\right)^\circ ,$$

if we set $V_n = H_n^\circ$. (P_ϵ) is immediate from the bipolar theorem. (Q_α),(Q_β),(Q_γ) are based on the fact that the families of (α')-, (β')-, and (γ')-sets are saturated,

respectively (cf. [50], p.141). (Q_δ) and (Q_ϵ) are immediate. From this it is also clear that analogous propositions $(Q_{\beta\delta})$, $(Q_{\beta\epsilon})$, $(Q_{\gamma\delta})$ and $(Q_{\gamma\epsilon})$ are valid, where the conjunctions of two items from the (x')- resp.(x'')-list occur.
By considering (γ), $(\gamma\delta)$, $(\gamma\epsilon)$, (β), $(\beta\delta)$ and $(\beta\epsilon)$ combined with (α), we obtain

1.2.1 Theorem *Let E be an lcs*

(i) E is barrelled iff each $\sigma(E', E)$-bounded subset of E' is equicontinuous.

(ii) E is infrabarrelled iff each $\sigma(E', E)$-bounded union of countably many equicontinuous subsets of E' is equicontinuous.

(iii) E is ω-barrelled iff each $\sigma(E', E)$-bounded sequence in E' is equicontinuous.

(iv) E is evaluable iff each $\beta(E', E)$-bounded subset of E' is equicontinuous.

(v) E is infraevaluable iff each $\beta(E', E)$-bounded union of countably many equicontinuous subsets of E' is equicontinuous.

(vi) E is ω-evaluable iff each $\beta(E', E)$-bounded sequence in E' is equicontinuous.

■

1.3 Strongly Bounded Sets in ω-Barrelled Spaces and their Duals

1.3.1 Lemma *If, in an lcs E, every $\sigma(E', E)$-bounded sequence is $\beta(E', E)$-bounded, then every $\sigma(E', E)$-bounded subset of E' is $\beta(E', E)$-bounded.*

Proof. Suppose there exists $B \subseteq E'$ which is $\sigma(E', E)$- bounded but not $\beta(E', E)$-bounded. Then there exists a bounded set $F \subseteq E'$ such that B is not uniformly bounded on F. Thus, for every $n \in \mathbf{N}$ we can find $f_n \in B$ and $x_n \in F$ with $| f_n(x_n) | > n$. The sequence $\{f_n \mid n \in \mathbf{N}\}$ provides the desired contradiction. ■

1.3.2 Proposition *Let E be ω-barrelled and $B \subseteq E'$. Then B is $\sigma(E', E)$-bounded if and only if B is $\beta(E', E)$-bounded. Equivalently, every barrel in E is bornivorous.*

Proof. Since $\beta(E', E) \geq \sigma(E', E)$, the sufficiency of the condition is clear.
Conversely, assume that B is $\sigma(E', E)$-bounded but not $\beta(E', E)$-bounded. By 1.3.1, there exists a $\sigma(E', E)$- bounded sequence $(f_n)_n$ which is not $\beta(E', E)$-bounded. But

since E is ω-barrelled, $(f_n)_n$ is equicontinuous and therefore $\beta(E', E)$-bounded (by [23],p.211,Prop.1). ∎

1.3.3 Corollary *If E is ω-barrelled, the bounded sets in E' are the same for all locally convex topologies τ on E' with $\sigma(E', E) \leq \tau \leq \beta(E', E)$.* ∎

1.3.4 Proposition *Every bounded subset X of an ω-barrelled space E is $\beta(E, E')$-bounded.*

Proof. If X is bounded in E and $B \subseteq E'$ is $\sigma(E', E)$- bounded then by 1.3.2 B is $\beta(E', E)$-bounded and therefore B is uniformly bounded on X. That is, X is uniformly bounded on B. Hence X is $\beta(E, E')$- bounded. ∎

1.3.5 Corollary *If E is ω-barrelled, the bounded sets in E are the same for each locally convex topology τ with $\sigma(E, E') \leq \tau \leq \beta(E, E')$.* ∎

1.4 (Bounded-)Absorbent Sequences

1.4.1 Definition *A sequence $(A_n)_n$ of subsets of an lcs E is an* absorbent sequence *if*

(i) Each A_n is absolutely convex,

(ii) $A_n \subseteq A_{n+1}$ for each $n \in \mathbf{N}$,

(iii) $\bigcup_{n=1}^{\infty} A_n$ is absorbing.

If, in addition to this

(iv) Each bounded set $B \subseteq E$ is absorbed by some A_n,

the sequence is called bounded-absorbent.

Obviously, (bounded-) absorbent sequences of closed sets are a generalized version of (bornivorous) barrels. Not surprisingly, several theorems dealing with barrels have their counterparts for absorbent sequences of closed sets: Compare the assertions of the following theorem with 1.3.2, (P_β), (P_γ) and the fact that, in an lcs E, any barrel (being a strong neighborhood of 0) absorbs every strongly bounded subset of E.

1.4.2 Theorem *Let $(F_n)_n$ be an increasing sequence of closed absolutely convex subsets of an lcs E.*

(*i*) Dual characterization

(*a*) *$(F_n)_n$ is absorbent if and only if each sequence $(f_n)_n$ with $f_n \in F_n^\circ$ for each n is $\sigma(E', E)$-bounded.*

(*b*) *$(F_n)_n$ is bounded-absorbent if and only if each sequence $(f_n)_n$ with $f_n \in F_n^\circ$ for each n is $\beta(E', E)$-bounded.*

(*ii*) *If $(F_n)_n$ is absorbent, then every strongly bounded subset of E is absorbed by some F_n.*

Proof.
(i)(a)($\Rightarrow$) Let $f_n \in F_n^\circ$, $x \in E$. There exists n_o such that $x \in n_o F_n$ for each $n \geq n_o$, hence $| f_n(x) | \leq n_o$ for $n \geq n_o$.
($\Leftarrow$) Conversely, assume that $(F_n)_n$ is not absorbent. Then there exists $x \in E$ such that $x \notin nF_n$ for each n. By the bipolar theorem we can choose $f_n \in F_n^\circ$ with $| f_n(x) | > n$.
(i)(b)($\Rightarrow$) Let $f_n \in F_n^\circ$, B a bounded subset of E. As in (a), there exists some n_o such that $\sup_{x \in B} | f_n(x) | \leq n_o$ for $n \geq n_o$. For $n < n_o$, $\sup_{x \in B} | f_n(x) | < \infty$ because B is bounded. Thus $\sup_{x \in B, n \in \mathbf{N}} | f_n(x) | < \infty$.
($\Leftarrow$) Assume that $(F_n)_n$ is not bounded-absorbent. Then there exist $x_n \in B \setminus nF_n$ for each n for some bounded subset B of E. Choose $f_n \in F_n^\circ$ with $| f_n(x_n) | > n$. Then $(f_n)_n$ is not $\beta(E', E)$-bounded since $\sup_{x \in B, n \in \mathbf{N}} | f_n(x) | = \infty$.
(ii) Let $(F_n)_n$ be absorbent and B a strongly bounded subset of E. Suppose that $B \not\subseteq nF_n$ for each n. Proceeding as in the second part of (i)(b) we obtain $x_n \in B \setminus nF_n$ and $f_n \in F_n^\circ$ satisfying $| f_n(x_n) | > n$ for each n. Since $(f_n)_n$ is $\sigma(E', E)$-bounded by (i)(a) we arrive at a contradiction. ■

1.4.3 Corollary *In an ω-barrelled space E, any absorbent sequence of closed sets is bounded absorbent.*

Proof. This is immediate from 1.3.4 and 1.4.2,(ii) or also from 1.3.2 and 1.4.2,(i)■

In our terminology, a (DF)-space is an infraevaluable lcs admitting a fundamental sequence of bounded sets (cf.[21],p.165).

1.4.4 Corollary *If the infrabarrelled space E possesses a countable family $(B_n)_n$ of bounded sets such that $\bigcup_{n \in \mathbf{N}} B_n$ is absorbing then E is a (DF)-space.*

Proof. For $n \in \mathbf{N}$ define F_n to be the closed absolutely convex hull of $\bigcup_{k=1}^{n} B_k$.

Then $(F_n)_n$ is an absorbent sequence of closed sets in E which is even bounded absorbent by 1.4.3. If we set $G_n := nF_n$, the sequence $(G_n)_n$ is a fundamental system of bounded sets in E. ■

1.4.5 Definition *Let A be a subset of a vector space V. The* algebraic closure *of A is the set $c(A) := \{y \mid \exists x \in A \text{ with } [x,y) \subseteq A\}$, where $[x,y) := \{\lambda y + (1-\lambda)x \mid 0 \le \lambda < 1\}$.* (cf.[28],p.180)

1.4.6 Lemma *If A is absolutely convex, then $c(A) = \bigcap_{\epsilon>0}(1+\epsilon)A$.*

Proof. Let $y \in c(A)$ and $\epsilon > 0$. Then, with $\lambda := \frac{2}{2+\epsilon}$ we have $\lambda y + (1-\lambda)x \in A$ for an arbitrary $x \in A$. Therefore $\lambda y \in (1-\lambda)A + A = (2-\lambda)A$, so that $y \in \frac{2-\lambda}{\lambda}A = (1+\epsilon)A$.
Conversely, assume that $y \in (1+\epsilon)A$ for each $\epsilon > 0$. That is, $(1+\epsilon)^{-1}y \in A$ for each $\epsilon > 0$. Consequently, $\lambda y \in A$ for $0 \le \lambda < 1$, i.e. $[0,y) \subseteq A$ and $y \in c(A)$. ■

1.4.7 Theorem *Let L be a subspace of an lcs E and $(A_n)_n$ an absorbent (resp. bounded-absorbent) sequence in L. If L is ω-barrelled (resp.ω-evaluable),then*

$$\overline{\bigcup_{n\in\mathbf{N}} A_n} = c\left(\bigcup_{n\in\mathbf{N}} \overline{A_n}\right),$$

where the closures are taken in E.

Proof. Clearly, for each $A \subseteq E$, $c(A) \subseteq \overline{A}$. Hence $c\left(\bigcup_{n\in\mathbf{N}} \overline{A_n}\right) \subseteq c\left(\overline{\bigcup_{n\in\mathbf{N}} A_n}\right) \subseteq \overline{\bigcup_{n\in\mathbf{N}} A_n}$.
As for the converse, by 1.4.6 it is sufficient to prove that

$$\overline{\bigcup_{n\in\mathbf{N}} A_n} \subseteq (1+\epsilon)\bigcup_{n\in\mathbf{N}} \overline{A_n} \quad \forall\epsilon > 0.$$

Without loss of generality we may assume that L is dense in E. Suppose that for some $\epsilon > 0$, $x \notin (1+\epsilon)\bigcup_{n\in\mathbf{N}} \overline{A_n}$. Then for each n there exists some $f_n \in E'$ with $f_n(x) = 1+\epsilon$ and $f_n \in A_n^\circ$. By 1.4.2 (i) it follows that the sequence $(f_n)_n$ is bounded in $(L', \sigma(L',L))$ or in $(L', \beta(L',L))$, according as $(A_n)_n$ is absorbent or bounded-absorbent. By 1.2.1, (iii) and (vi), $(f_n)_n$ is equicontinuous on L and hence equicontinuous on E because $\overline{L} = E$. The theorem of Alaoglu-Bourbaki now yields that $(f_n)_n$ is relatively $\sigma(E',E)$-compact. Therefore there exists a filter $\mathcal{F}$ on E' finer than the elementary filter $\mathcal{G}$ of $(f_n)_n$ such that $\mathcal{F}$ converges to some f in $(E', \sigma(E',E))$. In particular, $\mathcal{F}(x) \to f(x)$. Now $\mathcal{F}(x) \supseteq \mathcal{G}(x) \to 1+\epsilon$, hence

$f(x) = 1+\epsilon$. For each $m \in \mathbf{N}$ we have $A_k \supseteq A_m$ for all $k \geq m$, so that $A_k^\circ \subseteq A_m^\circ$ for all $k \geq m$. Thus, since f is an adherent point of $\mathcal{G}$: $f \in \overline{\{f_k \mid k \geq m\}}^{\sigma(E',E)} \subseteq A_m^\circ$ for each $m \in \mathbf{N}$, or equivalently $f \in \bigcap_{n=1}^\infty A_n^\circ = (\bigcup_{n=1}^\infty A_n)^\circ$. Finally, since $f(x) = 1+\epsilon$ we conclude that $x \notin \{f\}^\circ \supseteq (\bigcup_{n=1}^\infty A_n)^{\circ\circ} = \overline{(\bigcup_{n=1}^\infty A_n)}$ by the bipolar theorem. ■

We are now going to prove some results concerning neighborhoods of 0 in dense subspaces of lcs's which will turn out to be most valuable in simplifying proofs in this and the following chapters.

1.4.8 Lemma *Let M be a dense subset of a toplogical space E.*

(i) If U is an open subset of E, then $U \subseteq \overline{U \cap M}$.

(ii) If $x \in M$ and V is a neighborhood of x in M, then $\overline{V}$ is a neighborhood of x in E.

Proof. (i) Let $y \in U$ and let W be any neighborhood of y. Since $y \in \overline{M}$, $\emptyset \neq (W \cap U) \cap M = W \cap (U \cap M)$. Therefore, $y \in \overline{U \cap M}$.
(ii) Choose an open neighborhood U of x in E such that $U \cap M \subseteq V$. Then $\overline{U \cap M} \subseteq \overline{V}$ and by (i), $\overline{V}$ is a neighborhood of x in E. ■

1.4.9 Corollary *Let M be a dense subspace of a tvs E and let W be a closed, absolutely convex neighborhood of 0 in E. Then $W = \overline{W \cap M}$.*

Proof. 1.4.8 (i) yields $int(W) \subseteq \overline{W \cap M} \subseteq W$. From $\lambda W + (1-\lambda)W \subseteq W$ for $0 \leq \lambda < 1$ it follows that $\bigcup_{0 \leq \lambda < 1} \lambda W \subseteq int(W)$. This in turn implies $W \subseteq \overline{\bigcup_{0 \leq \lambda < 1} \lambda W} \subseteq \overline{int(W)} \subseteq W$. Altogether, we have $W \subseteq \overline{int(W)} \subseteq \overline{W \cap M} \subseteq W$. ■

1.4.10 Corollary *Let $(A_n)_n$ be an absorbent (resp. bounded-absorbent) sequence in an ω-barrelled (resp. ω-evaluable) space L. If $\mathcal{F}$ is a Cauchy filter on $\bigcup_{n \in \mathbf{N}} A_n$ and $\mathcal{G}$ is the filter with basis $\mathcal{B} = \{M + U \mid M \in \mathcal{F}, U$ is a neighborhood of 0 in $L\}$, then for each $\epsilon > 0$ $\mathcal{G}$ induces a Cauchy filter on some $(1+\epsilon)A_{n_o}$.*

Proof. If E denotes the completion of L then $\mathcal{F}$ converges to some $x \in E$. Fix $\epsilon > 0$. By 1.4.7, $x \in (1+\epsilon)\overline{A_{n_o}}$ for some $n_o \in \mathbf{N}$. Choose $M \in \mathcal{F}$ and U a closed, absolutely convex neighborhood of 0 in L. Since x is an adherent point of $\mathcal{F}$ and $\overline{U}$ is a neighborhood of 0 in E (cf.1.4.8 (ii)), $x \in M + \frac{1}{2}\overline{U}$. Furthermore, $(x + \frac{1}{2}\overline{U}) \cap (1+\epsilon)A_{n_o} \neq \emptyset$. Choose $y = x + \frac{1}{2}\overline{u}$ (where $\overline{u} \in \overline{U}$) in this last intersection and $\overline{v} \in \overline{U}, m \in M$ such that $x = m + \frac{1}{2}\overline{v}$. Then $y = m + \frac{1}{2}\overline{u} + \frac{1}{2}\overline{v} \in M + \overline{U}$, so that $(M + \overline{U}) \cap (1+\epsilon)A_{n_o} \neq \emptyset$. Moreover, $y = m + \frac{1}{2}\overline{u} + \frac{1}{2}\overline{v} \in (1+\epsilon)A_{n_o} \subseteq L$ which, together with $m \in L$, gives $\frac{1}{2}\overline{u}+\frac{1}{2}\overline{v} \in L \cap \overline{U} = U$. Thus even $(M+U)\cap(1+\epsilon)A_{n_o} \neq \emptyset$, completing the proof. ■

1.4.11 Remark One can also deduce 1.4.7 from 1.4.10:

Proof. If $x \in \overline{\bigcup_{n\in\mathbf{N}} A_n}$ then $\mathcal{F}$, the trace of the neighborhood filter of x on L, is a Cauchy filter on $\bigcup_{n\in N} A_n$. Hence $\mathcal{G} = \mathcal{F}$ induces a filter on some $(1+\epsilon)A_{n_o}$, so that $x \in (1+\epsilon)\overline{A_{n_o}}$. ■

1.4.12 Corollary *If E is ω-barrelled (resp. ω-evaluable), then the union of each absorbent (resp. bounded-absorbent) sequence of closed sets contains a barrel (resp. bornivorous barrel).*

Proof. 1.4.7 (with L=E, $\epsilon = 1$) yields: $\frac{1}{2}\overline{\bigcup_{n\in\mathbf{N}} A_n} \subseteq \bigcup_{n\in\mathbf{N}} A_n$. ■

1.4.13 Corollary *Let E be ω-barrelled (resp. ω- evaluable). If there exists an absorbent (resp. bounded-absorbent) sequence $(A_n)_n$ of complete subsets of E, then E is complete.*

Proof. $(nA_n)_n$ is absorbent (resp. bounded absorbent) and $E = \bigcup_{n=1}^{\infty} nA_n$. If we take the closures in the completion $\widetilde{E}$ of E, 1.4.7 yields:

$$\widetilde{E} = \overline{E} = \frac{1}{2}\overline{\bigcup_{n=1}^{\infty} nA_n} \subseteq \bigcup_{n=1}^{\infty} n\overline{A_n} = \bigcup_{n=1}^{\infty} nA_n = E$$

■

1.4.14 Lemma *Let (E,τ) be a tvs. If the absolutely convex subset A of E has nonempty interior then it is a neighborhood of 0*

Proof. The open subset $\frac{1}{2}int(A) - \frac{1}{2}int(A)$ of A contains 0. ■

1.4.15 Corollary *Let E be an ω-barrelled (resp. ω-evaluable) space whose completion $\widetilde{E}$ is a Baire space. If $(F_n)_n$ is an absorbent (resp. bounded-absorbent) sequence of closed subsets of E, then one of the F_n is a neighborhood of 0.*

Proof. $(nF_n)_n$ is a sequence which is of the same type as $(F_n)_n$ and covers E. By the proof of 1.4.13, $\widetilde{E} = \bigcup_{n=1}^{\infty} n\overline{F_n}$, where the closures are taken in $\widetilde{E}$. $\widetilde{E}$ is Baire, so that some $\overline{F_n}$ has nonempty interior and therefore is a neighborhood of 0 by 1.4.14. Hence $F_n = \overline{F_n} \cap E$ is a neighborhood of 0 in E. ■

1.4.16 Remark In the terminology of chapter 5, this corollary can be restated as follows: An ω-barrelled space is Baire-like if its completion is a Baire space (in fact, as inspection of the proof shows, Baire-like is sufficient, cf. 6.2.14).

1.4.17 Corollary *Let E be an ω-barrelled space whose completion is a Baire space. If M is a closed, countable-codimensional subspace of E, then its codimension is necessarily finite.*

Proof. Suppose $codim(M) = \aleph_0$ and let $\{x_n \mid n \in \mathbf{N}\}$ be a Hamel basis of some algebraic complement of M. For each $n \in \mathbf{N}$ set $F_n := sp(M \cup \{x_k \mid 1 \leq k \leq n\})$. Then $(F_n)_n$ satisfies the hypothesis of 1.4.15, so that some F_n is a neighborhood of 0 in E. But then $F_n = E$, a contradiction. ∎

1.4.18 Definition *A tvs E is called* locally bounded *if it possesses a bounded neighborhood of zero.*

By [50], II.2.1, every locally bounded lcs is normable.

1.4.19 Corollary *Let E be an ω-barrelled space whose completion is a Baire space. If there exists a sequence $(B_n)_n$ of bounded subsets of E such that $\bigcup_{n=1}^{\infty} B_n$ is absorbing, then E is normable.*

Proof. We show that E is locally bounded: Let A_n denote the closed, absolutely convex hull of $\bigcup_{k=1}^{n} B_k$. Then $(A_n)_n$ is an absorbent sequence in E. By 1.4.15, some A_n is a bounded neighborhood of 0. ∎

1.5 Tools for Localization

1.5.1 Theorem *Let $(W_n)_n$ be an absorbent sequence in the barrelled space E with $E = \bigcup_{n=1}^{\infty} W_n$. If U is an absorbing absolutely convex set such that $U \cap W_n$ is closed in W_n for each n, then U is a neighborhood of 0 in E.*

Proof. The sequence $(A_n)_n$ with $A_n = U \cap W_n$ for each $n \in \mathbf{N}$ is absorbent. Let x be an arbitrary element of the closure of $U = \bigcup_{n=1}^{\infty} A_n$ and $\mathcal{F}$ the trace of the neighborhood filter of x on U. Then for some n_o the filter $\mathcal{G}$ described in 1.4.10 induces a Cauchy filter $\mathcal{G}'$ on $2A_{n_o}$. Obviously n_o can be chosen in such a way that $x \in 2W_{n_o}$. $\mathcal{G}$ converges to x since $\mathcal{F}$ does ($\mathcal{F} \supseteq \mathcal{G}$ and both are Cauchy filters). Hence $\mathcal{G}'$ converges to x and since $2A_{n_o}$ is closed in $2W_{n_o}$ it follows that $x \in 2A_{n_o} \subseteq 2U$, i.e. $\overline{U} \subseteq 2U$. $\overline{U}$ is a barrel in E, so that 2U is a neighborhood of the origin. ∎

1.5.2 Corollary *Let $(E_n)_n$ be an increasing sequence of subspaces of the barrelled space E whose union is E. If U is a subset of E such that $U \cap E_n$ is a barrel in E_n*

for each n, then U is a neighborhood of 0 in E.

Proof. It is easily checked that U is absolutely convex and absorbing. Now take $E_n = W_n$ in 1.5.1. ■

1.5.3 Theorem *Every countable-codimensional subspace E_1 of a barrelled space E is barrelled.*

Proof. The finite-codimensional case is settled by 1.1.1. Now let $(x_n)_n$ be a sequence in E such that $E_1 \cup \{x_n \mid n \in \mathbf{N}\}$ generates E. Denote by E_n the linear span of E_1 and $\{x_k \mid 1 \leq k \leq n\}$ and let U_1 be a barrel in E_1. By 1.1.2 we can inductively define a sequence $(U_n)_n$ such that $U_{n+1} \cap E_n = U_n$ and U_n is a barrel in E_n for each $n \in \mathbf{N}$. If $U = \bigcup_{n=1}^{\infty} U_n$, then $U \cap E_n = \left(\bigcup_{k=n+1}^{\infty} U_k\right) \cap E_n = U_n$ for each n. By 1.5.2, U is a neighborhood of 0 in E. Hence $U_1 = E_1 \cap U$ is a neighborhood of 0 in E_1. ■

1.5.4 Remark The corresponding result for evaluable spaces is false (see [54]). It is true, though, for finite codimensional subspaces (cf. [24], p.226).

1.5.5 Proposition *If $(A_n)_n$ is a bounded-absorbent sequence in a metrizable lcs E, then one of the A_n is a neighborhood of 0.*

Proof. Let $\{U_n \mid n \in \mathbf{N}\}$ be a neighborhood base of 0 in E and suppose no A_n is a neighborhood of zero. Then for each $n \in \mathbf{N}$, $U_n \not\subseteq nA_n$ and hence there exists some $x_n \in U_n \setminus nA_n$. Since $(x_n)_n$ converges to 0 it is bounded and thus it is absorbed by some A_{n_o}, say $x_n \in CA_{n_o}$ for each n. But this contradicts the fact that $CA_{n_o} \subseteq nA_n$ for n big enough. ■

1.5.6 Theorem *Let $(A_n)_n$ be an absorbent (resp. bounded-absorbent) sequence in an infrabarrelled (resp. infraevaluable) space E and let $(\lambda_n)_n \nearrow \infty$. Then any absolutely convex set U such that $U \cap \lambda_n A_n$ is a neighborhood of 0 in $\lambda_n A_n$ for each $n \in \mathbf{N}$ is a neighborhood of 0 in E.*

Proof. For each n, choose U_n an absolutely convex neighborhood of 0 in E such that $U_n \cap \lambda_n A_n \subseteq U$. We claim that $I := \bigcap_{n=1}^{\infty} \overline{(U \cap \lambda_n A_n) + U_n}$ is an infrabarrel (resp. bornivorous infrabarrel). Indeed, it is the intersection of a sequence of closed and absolutely convex neighborhoods of 0 in E.

It remains to show that I absorbs each element (resp. bounded subset) of E. We prove this explicitly only for a bounded set B: For some $n_o \in \mathbf{N}$ and some $\lambda > 0$ we have $B \subseteq \lambda_{n_o} A_{n_o}$ and $B \subseteq \lambda U_{n_o}$. Consequently, $B \subseteq \max(1, \lambda)(U_{n_o} \cap \lambda_{n_o} A_{n_o}) \subseteq$

$\max(1,\lambda)(U \cap \lambda_n A_n)$ for each $n \geq n_o$. If we fix μ such that $B \subseteq \mu U_n$ for all $n < n_o$ it follows that $B \subseteq \max(1,\lambda,\mu)I$.
By our assumption, I is a neighborhood of 0 in E. Let $x \in I$. Then for some n_o, $x \in \lambda_{n_o} A_{n_o}$. Furthermore, $I \subseteq \overline{(U \cap \lambda_{n_o} A_{n_o}) + U_{n_o}} \subseteq (U \cap \lambda_{n_o} A_{n_o}) + 2U_{n_o}$, so that $x = y + z$, where $y \in U \cap \lambda_{n_o} A_{n_o}$ and $z \in 2U_{n_o}$. Since $x \in \lambda_{n_o} A_{n_o}$, we get $z \in 2(U_{n_o} \cap \lambda_{n_o} A_{n_o}) \subseteq 2U$, hence $x \in 3U$. Consequently, $I \subseteq 3U$ and the proof is complete. ■

1.5.7 Corollary *Let $(A_n)_n$ be an absorbent (resp. bounded-absorbent) sequence in an infrabarrelled (resp. infraevaluable) space E with $E = \bigcup_{n=1}^{\infty} A_n$. Then any absolutely convex set U such that $U \cap A_n$ is a neighborhood of 0 in A_n for each n is a neighborhood of 0 in E.*

Proof. A slight modification of the proof of 1.5.6. ■

1.5.8 Corollary *Let $(A_n)_n$ and E be as in 1.5.7. If f is a linear mapping from E into any lcs F, then f is continuous if and only if its restriction to each A_n is continuous.* ■

1.5.9 Corollary *Let $(E_n)_n$ be an absorbent (resp. bounded-absorbent) sequence of subspaces of the infrabarrelled (resp. infraevaluable) space (E,τ). Then (E,τ) is the strict inductive limit (cf. 7.1.1) of the sequence $((E_n, \tau|_{E_n}))_n$.*

Proof. Let $(E,\sigma) = \underrightarrow{\lim}(E_n, \tau|_{E_n})$. Then clearly $\tau \leq \sigma$.
Conversely, let V be an absolutely convex σ-neighborhood of 0 in E. Then $V \cap E_n$ is a neighborhood of 0 in E_n for each n and by 1.5.7, V is a τ- neighborhood of 0 in E. ■

1.5.10 Theorem *If L is a separable subspace of the ω-barrelled (resp. ω-evaluable) space E, then any barrel (resp. bornivorous barrel) B in E induces a neighborhood of 0 in L.*

Proof. $L \setminus B$ is open in L and therefore separable. Let $D = \{x_n \mid n \in \mathbf{N}\}$ be a dense subset of $L \setminus B$. By the Hahn-Banach theorem we can choose a sequence $(f_n)_n$ in E' such that $f_n \in B^\circ$ and $f_n(x_n) > 1$ for each $n \in \mathbf{N}$. Set $U := \{f_n \mid n \in \mathbf{N}\}^\circ = \bigcap_{n=1}^{\infty}\{f_n\}^\circ$. Then $B \subseteq U$, so that U is absorbing and thus is an ω-barrel (resp. bornivorous ω-barrel). By our assumption on E, U is a neighborhood of 0 in E and therefore $int(U) \neq \emptyset$. Now $x_n \notin U$ implies $D \subseteq L \setminus U \subseteq L \setminus int(U)$, so that $L \setminus B \subseteq L \cap \overline{D} \subseteq L \setminus int(U)$. Hence $int(U) \cap L \subseteq B$ and $B \cap L$ is a neighborhood of 0 in L. ■

1.5.11 Corollary *If E is separable and ω-barrelled (resp. ω-evaluable), then it is barrelled (resp. evaluable).* ■

1.5.12 Corollary *If (E,τ) is ω-barrelled (resp. ω-evaluable) and if σ is that locally convex topology on E for which a fundamental system of neighborhoods of 0 is given by the family of all barrels (resp. bornivorous barrels) in E, then a sequence converges in (E,τ) if and only if it converges in (E,σ).*

Proof. If $x \in E$ and $(x_n)_n$ is a sequence in E, then $L := span(\{x_n \mid n \in \mathbf{N}\} \cup \{x\})$ is a separable subspace of E. Hence the topologies induced by τ and σ are equivalent on L. Thus $(x_n)_n$ converges to x in (E,τ) if and only if it converges to x in (E,σ).■

In 1.5.12, σ is just the strong topology on E or the topology of uniform convergence on the strongly bounded subsets of E', respectively (cf. 1.2).

1.5.13 Corollary *If in the infrabarrelled (resp. infraevaluable) space E there exists an absorbent (resp. bounded-absorbent) sequence $(A_n)_n$ of metrizable subsets, then E is barrelled (resp. evaluable).*

Proof. For each n, nA_n is metrizable. Let B be a barrel (resp. bornivorous barrel) and denote by p_B its gauge. By 1.5.12, any convergent sequence in E converges for p_B. But then $B \cap nA_n$ is a neighborhood of 0 in nA_n: Otherwise, by means of a countable fundamental system of neighborhoods of 0 in nA_n, we could define a sequence $(x_m)_m$ in nA_n converging to 0 such that no x_m is in B, contradicting $p_B(x_m) \to 0$. Hence 1.5.6 yields the result. ■

In particular, every metrizable infrabarrelled (resp. infraevaluable) space E is barrelled (resp. evaluable).

1.6 Characterization by Locally Convex Topologies

First of all, we are going to concern ourselves with four more generalizations of barrelledness that will turn out to be closely related to questions arising in the context of Mackey spaces. Recall that a subset A of a Hausdorff topological space X is called *countably compact* if every sequence in A has a cluster point in A (cf.[50], p.185).

1.6.1 Definition *An lcs E is said to have*

Property (C), *if every $\sigma(E',E)$-bounded subset of E' is relatively $\sigma(E',E)$-countably* *compact.*

Property (S), *if E' is $\sigma(E',E)$-sequentially complete.*

A c_o-barrel *in E is the polar of a $\sigma(E',E)$-null sequence in E'.*

E is c_o-barrelled *(resp.* c_o-evaluable*) if each c_o-barrel (resp. bornivorous c_o-barrel) in E is a neighborhood of 0.*

For an arbitrary dual system $\langle E,F\rangle$, denote by $\mu(E,F)$ the Mackey topology associated with $\langle E,F\rangle$.

1.6.2 Remarks

(i) An lcs E is c_o-barrelled (resp. c_o-evaluable) if and only if each $\sigma(E',E)$-null-sequence (resp.$\beta(E',E)$-bounded $\sigma(E',E)$-null-sequence) in E' is equicontinuous.(cf. 1.2)

(ii) By the theorem of Alaoglu-Bourbaki, every ω-barrelled lcs has property (C).

(iii) Property (C) implies property (S) since each Cauchy sequence with an adherent point converges.

(iv) Every ω-barrelled space is c_o-barrelled as can be seen directly from the definitions.

The properties of being (infra-, ω-, c_o-) barrelled or -evaluable, respectively, also admit descriptions based on the comparison of certain locally convex topologies. The technical details are as follows (cf. e.g. [50], p.79f or [23], p.185f):

Let $\langle F,G\rangle$ be a dual system and Υ a family of subsets of G possessing the following properties:

(i) For $A_1, A_2 \in \Upsilon$ there exists $A_3 \in \Upsilon$ satisfying $A_1 \cup A_2 \subseteq A_3$.

(ii) For $A \in \Upsilon$ and $0 \neq \lambda \in \mathbf{K}$, $\lambda A \in \Upsilon$.

(iii) Every $A \in \Upsilon$ is $\sigma(G,F)$-bounded.

(iv) $\bigcup\{A \mid A \in \Upsilon\}$ spans a $\sigma(G,F)$-dense subspace of G.

Then $\Upsilon^\circ := \{A^\circ \mid A \in \Upsilon\}$ forms a fundamental system of neighborhoods of 0 for a Hausdorff locally convex topology on F, the *topology of uniform convergence on the sets $A \in \Upsilon$*, which we denote by $\mathcal{T}_\Upsilon$. The *saturated hull* $\overline{\Upsilon}$ of Υ consists of all subsets of sets $\overline{\Gamma}^\sigma A$, $A \in \Upsilon$; $\mathcal{T}_\Upsilon$ is equal to $\mathcal{T}_{\overline{\Upsilon}}$.

All the families used in the sequel to define topologies on a member of a dual system indeed satisfy (i)-(iv); we omit the easy proofs.

1.6.3 Proposition *Let $\langle F, G\rangle$ be a dual system and Υ_1, Υ_2 families of subsets of G satisfying (i)-(iv).*

(a) If $\Upsilon_1 \subseteq \Upsilon_2$, then $\mathcal{T}_{\Upsilon_1} \leq \mathcal{T}_{\Upsilon_2}$.

(b) Assume, in addition, that Υ_2 contains all $\sigma(G, F)$-closed absolutely convex hulls and all subsets of all of its members. Then we also have:
If $\mathcal{T}_{\Upsilon_1} \leq \mathcal{T}_{\Upsilon_2}$, then $\Upsilon_1 \subseteq \Upsilon_2$.

Proof. (a) is immediate.
(b) Let $A_1 \in \Upsilon_1$. A_1°, being a neighborhood of 0 with respect to $\mathcal{T}_{\Upsilon_1}$, contains a set A_2°, $A_2 \in \Upsilon_2$. It follows that $A_1 \subseteq A_1^{\circ\circ} \subseteq A_2^{\circ\circ} = \overline{\Gamma}^{\sigma} A_2 \in \Upsilon_2$ and $A_1 \in \Upsilon_2$. ■

1.6.4 Definition *For a given duality $\langle E, F\rangle$, let $\beta(E,F)$, $\nu(E,F)$, $\nu_o(E,F)$ denote the topologies (on E) of uniform convergence on the $\sigma(F,E)$-bounded subsets of F which are arbitrary, resp. countable, resp. constitute a $\sigma(F,E)$-null-sequence.*
Define $\beta^*(E,F)$, $\nu^*(E,F)$, $\nu_o^*(E,F)$ in an analogous way, with '$\sigma(F,E)$-bounded' replaced by '$\beta(F,E)$-bounded'.
For a given lcs (E,τ), let $\eta(E,\tau,E')$ (resp. $\eta^*(E,\tau,E')$) denote the topology (on E) of uniform convergence on the $\sigma(E',E)$-bounded (resp. $\beta(E',E)$-bounded) subsets of E' which are of the form $\bigcup_{n=1}^{\infty} H_n$, where the H_n are equicontinuous with respect to τ.

From these definitions (and from [50], IV.5.1), we have the following relations between the topologies introduced above:

$$\begin{array}{ccccc}
\nu_o(E,F) & \leq & \nu(E,F) & \leq & \beta(E,F) \\
\text{\rotatebox{90}{$\leq$}} & & \text{\rotatebox{90}{$\leq$}} & & \text{\rotatebox{90}{$\leq$}} \\
\nu_o^*(E,F) & \leq & \nu^*(E,F) & \leq & \beta^*(E,F) \\
\text{\rotatebox{90}{$\leq$}} & & & & \text{\rotatebox{90}{$\leq$}} \\
\sigma(E,F) & & \leq & & \mu(E,F)
\end{array}$$

and

$$
\begin{array}{ccccccc}
\nu_o(E,E') & \leq & \nu(E,E') & \leq & \eta(E,\tau,E') & \leq & \beta(E,E') \\
\geq & & \geq & & \geq & & \geq \\
\nu_o^*(E,E') & \leq & \nu^*(E,E') & \leq & \eta^*(E,\tau,E') & \leq & \beta^*(E,E') \\
\geq & & & & \geq & & \geq \\
\sigma(E,E') & & \leq & & \tau & \leq & \mu(E,E')
\end{array}
$$

Using the dual characterizations given in 1.2, the following proposition is immediate from 1.6.3:

1.6.5 Proposition *An lcs (E,τ) is barrelled [infrabarrelled, ω-barrelled, c_o-barrelled, evaluable, infraevaluable, ω-evaluable, c_o-evaluable] if and only if β [$\eta, \nu, \nu_o, \beta^*, \eta^*, \nu^*, \nu_o^*$; add '$(E,E')$' resp. '$(E,\tau,E')$' in each case] is coarser than its given topology τ* ■

Observe that $\beta, \eta, \beta^*, \eta^*$ are finer than τ, in any case.

1.6.6 Proposition *For a dual system $\langle E,F\rangle$, let ρ be one of the topologies $\beta(E,F)$, $\nu(E,F)$, $\nu_o(E,F)$, $\beta^*(E,F)$, $\nu^*(E,F)$, $\nu_o^*(E,F)$. If ρ is compatible with $\langle E,F\rangle$ and ξ is a locally convex topology on E satisfying $\rho \leq \xi \leq \mu(E,F)$, then (E,ξ) is barrelled [ω-barrelled, c_o-barrelled, evaluable, ω-evaluable, c_o-evaluable].*

Proof. For example, let $\rho = \nu(E,F)$ and $\rho \leq \xi \leq \mu(E,F)$. Then $\nu(E,(E,\xi)') = \nu(E,F) = \rho \leq \xi$. By 1.6.5, (E,ξ) is ω-barrelled. ■

It appears natural to ask for an analogue of 1.6.6 for the properties of being infrabarrelled resp. infraevaluable. So let (E,τ) be an lcs, $E' = (E,\tau)'$ and assume that $\rho = \eta(E,\tau,E')$ [or $\rho = \eta^*(E,\tau,E')$] is compatible with $\langle E,E'\rangle$ and $\rho \leq \xi \leq \mu(E,E')$. From this we conclude $\eta(E,\xi,(E,\xi)') = \eta(E,\xi,E')$ and $\eta(E,\tau,E') = \rho \leq \xi$ [similarly for η^*]. To procede as in the proof of 1.6.6, we would need $\eta(E,\xi,E') \leq \eta(E,\tau,E')$. However, the obviously sufficient condition for this inequality, namely $\xi \leq \tau$, leads to $\tau \leq \eta(E,\tau,E') = \rho \leq \xi \leq \tau$, i.e. $\xi = \tau = \eta(E,\tau,E')$. To reach the desired conclusion that (E,ξ) is infrabarrelled, we would have to assume (E,τ) to be infrabarrelled and $\xi = \tau$, so nothing would be gained this way. The reason for this failure, of course, lies in the fact that η and η^* do not only depend on the duality $\langle E,E'\rangle$, but also on the topologies τ and ξ which do not necessarily determine the same equicontinuous sets in E'.

1.6.7 Proposition *If $(E,\mu(E,E'))$ is quasi-complete, then $\nu_o(E',E) \leq \mu(E',E)$.*

Proof. For an arbitrary $\sigma(E,E')$-null sequence $(x_n)_n$ in E let B denote the closed absolutely convex hull of $\{x_n \mid n \in \mathbf{N}\} \cup \{0\}$. B is $\mu(E, E')$-complete since it is closed and bounded in $(E, \mu(E, E'))$. Now $\{x_n \mid n \in \mathbf{N}\} \cup \{0\}$ is $\sigma(E, E')$-compact, so that by Krein's theorem (cf. [50], IV.11.4) B is $\sigma(E, E')$-compact. Hence B°, a typical $\nu_o(E', E)$-neighborhood of 0, is a $\mu(E', E)$-neighborhood of 0. ■

1.6.8 Corollary *For any Banach space E, the dual E' is c_o-barrelled when equipped with any locally convex topology ξ such that $\nu_o(E', E) \leq \xi \leq \mu(E', E)$.*

Proof. E is a complete Mackey space. Apply 1.6.6 and 1.6.7. ■

1.7 Countable-Codimensional Subspaces

In this and the subsequent chapters we will repeatedly encounter situations in which important properties of lcs's are inherited by countable-codimensional subspaces (cf.1.5.3). Indeed, this behaviour appears to be a distinctive feature of many concepts we are dealing with in this treatise. Recall that, in our terminology, 'countable' always comprises both, the finite and the countably infinite case.

1.7.1 Proposition *Let E be an lcs with property (S). A linear functional f on E is continuous if and only if its restriction to any countable-codimensional subspace M of E is continuous.*

Proof. If $codim(M) < \infty$, the result is obvious, since then each algebraic complement of M in E is topological.
Otherwise let $\{x_n \mid n \in \mathbf{N}\}$ be a Hamel basis for some algebraic complement of M. Suppose f is a linear functional on E and $f|_M$ is continuous. Since M is a closed, finite codimensional subspace of $M_n := M \oplus sp(x_1, ..., x_n)$, the map $f|_{M_n}$ is a continuous extension of $f|_M$ to M_n which can itself be extended to some $g_n \in E'$. The sequence $(g_n)_n$ is $\sigma(E', E)$-convergent to f. Thus $f \in E'$ since E has property (S) ■

1.7.2 Proposition *Let (E, τ) be a Mackey space with property (S), M a closed subspace of countable codimension in E and N any algebraic complement of M. Then N carries the strongest locally convex topology and is a topological complement of M in E.*

Proof. Denote by P the projection of E onto N along M. Let V be any absorbing, absolutely convex subset of N. Let τ_o be that locally convex topology on E which has as a fundamental system of neighborhoods of 0 all sets of the form

$U \cap \epsilon(M+V) = U \cap (M+\epsilon V)$, where U is a τ-neighborhood of 0 and $\epsilon > 0$. Then $\tau_o \geq \tau$, so that $(E,\tau)' \subseteq (E,\tau_o)'$. Moreover, $\tau\mid_M = \tau_o\mid_M$. Hence by 1.7.1 we have: $f \in (E,\tau_o)' \Rightarrow f\mid_M \in (M,\tau_o\mid_M)' = (M,\tau\mid_M)' \Rightarrow f \in (E,\tau)'$, so that $(E,\tau)' = (E,\tau_o)'$. Since (E,τ) is a Mackey space, $\tau_o = \tau$. In particular, $M+V$ is a τ-neighborhood of the origin in N. That means that N carries the strongest locally convex topology. On the other hand, $P^{-1}(V) = M+V$ implies that P is continuous which proves the remaining part of our assertion. ■

1.7.3 Corollary *If E is countable-dimensional and barrelled, it carries the strongest locally convex topology.*

Proof. Choose $M = \{0\}$ in 1.7.2. ■

We are now going to examine inheritance properties of countable-codimensional subspaces of various kinds of 'almost barrelled' spaces. First of all, we need

1.7.4 Lemma *Let A be an absolutely convex closed subset of an lcs E with property (S). If sp(A) has countable codimension in E, then it is closed in E.*

Proof. We assume that $codim(sp(A)) = \aleph_0$. (The modifications for the finite-codimensional case are obvious).
Let x_1 be an arbitrary element of $E \setminus sp(A)$ and choose $x_2, x_3, \ldots$ in such a way that $(x_n)_n$ is a linearly independent sequence with $E = sp(\{x_n \mid n \in \mathbf{N}\}) \oplus sp(A)$. Since no nonzero scalar multiple of any x_n is in $A = A^{\circ\circ}$ it follows that A° is unbounded on each x_n. Let $0 < \epsilon < 1$.
Define $A_1 := A$ and $A_n := A_{n-1} + \{tx_n \mid |t| \leq 1\}$ for $n = 2, 3, \ldots$. Then each A_n is absolutely convex and closed, being the sum of a closed set and a compact set. Thus $A_n^{\circ\circ} = A_n$ for each n by the bipolar theorem. Moreover, $x_n \notin sp(A_n)$, so that we can choose a sequence $(f_n)_n$ in E' with $f_n \in 2^{-n}\epsilon A_n^\circ$, $f_1(x_1) = 2$ and $\sum_{k=1}^{n+1} f_k(x_{n+1}) = 0$ for $n = 1, 2, \ldots$.
Let $x \in E$. $E = \bigcup_{n=1}^\infty sp(A_n)$ and A_n is absorbing in $sp(A_n)$, so that for some $p \in \mathbf{N}$ and some $\delta > 0$ we have $\delta x \in A_p \subseteq A_{p+1} \subseteq \ldots$. Therefore

$$\sum_{k=1}^\infty |f_k(x)| = \frac{1}{\delta}\sum_{k=1}^\infty |f_k(\delta x)| \leq \frac{1}{\delta}\left(\sum_{k=1}^{p-1} |f_k(\delta x)| + \sum_{k=p}^\infty 2^{-k}\epsilon\right) < \infty.$$

Since E has property (S), $f_\epsilon(x) := \sum_{k=1}^{\infty} f_k(x)$ defines a continuous linear functional on E. If $x \in A_1$ then

$$| f_\epsilon(x) | \leq \sum_{k=1}^{\infty} | f_k(x) | \leq \sum_{k=1}^{\infty} 2^{-k}\epsilon = \epsilon$$

and

$$| f_\epsilon(x_{n+1}) | \leq | \sum_{k=1}^{n+1} f_k(x_{n+1}) | + \sum_{k=n+2}^{\infty} | f_k(x_{n+1}) | \leq 0 + \sum_{k=n+2}^{\infty} 2^{-k}\epsilon < \epsilon$$

for each $n \in \mathbf{N}$.
In addition to this,$| f_\epsilon(x_1) | \geq 2 - \sum_{k=2}^{\infty} | f_k(x_1) | > 2 - \epsilon > 1$. Let $g_\epsilon := (f_\epsilon(x_1))^{-1} f_\epsilon$. Then $g_\epsilon(x_1) = 1$, $| g_\epsilon(x) | < \epsilon$ for each $x \in A_1 = A$ and $| g_\epsilon(x_{n+1}) | < \epsilon$ for $n = 1, 2, \ldots$.
Let $(\epsilon_n)_n$ be a sequence in $]0,1[$ converging to 0.
Then h, defined as the unique linear extension to E of

$$x \mapsto \begin{cases} 0 & \text{if } x \in A_1 \\ 1 & \text{if } x = x_1 \\ 0 & \text{if } x = x_{n+1} \text{ for some } n = 1, 2, \ldots \end{cases}$$

is the pointwise limit of the sequence $(g_{\epsilon_n})_n$ and hence is continuous since E has property (S). It follows that $\overline{sp(A)} \subseteq h^{-1}(0)$ and that $x_1 \notin \overline{sp(A)}$. Since each $x \notin sp(A)$ can take the place of x_1, $\overline{sp(A)} \subseteq sp(A)$ and the proof is complete. ■

1.7.5 Proposition *Let E be a locally convex space with property (S) and let M be a dense, countable-codimensional subspace of E. Then each $\sigma(E', M)$-bounded subset B of E' is $\sigma(E', E)$-bounded.*

Proof. B° is a closed and absolutely convex subset of E. Since B is $\sigma(E', M)$-bounded, M is absorbed by B°. Now M is dense in E and $sp(B^\circ)$ is closed in E by 1.7.4. Thus $E = sp(B^\circ)$, i.e. B is $\sigma(E', E)$-bounded. ■

1.7.6 Proposition *Let M be a dense, countable-codimensional subspace of an lcs E. If E is ω-barrelled, has property (C) or has property (S), then M has the corresponding property.*

Proof. We identify E' canonically with M'.
Suppose E is ω-barrelled and let B be any countable $\sigma(E', M)$-bounded subset of E'. By 1.7.5, B is $\sigma(E', E)$-bounded, hence equicontinuous on E, hence equicontinuous on M. Thus M is ω-barrelled.

If E has property (C), choose an arbitrary infinite $\sigma(E', M)$-bounded subset B of E'. Again by 1.7.5, B is $\sigma(E', E)$-bounded. Hence every sequence $(f_n)_n$ in B has a $\sigma(E', E)$-cluster point f in E'. Since $\sigma(E', E) \geq \sigma(E', M)$, f is a $\sigma(E', M)$-cluster point of B and M, too, has property (C).
Finally, assume that E has property (S) and let $(f_n)_n$ be a $\sigma(E', M)$-Cauchy sequence in E'. The set $\{f_n \mid n \in \mathbf{N}\}$ is $\sigma(E', M)$-bounded and therefore $\sigma(E', E)$-bounded by 1.7.5. Let X be a countable subset of E such that $M \cup X$ spans E. Then for each $x \in X$, $(f_n(x))_n$ is a bounded sequence in $\mathbf{K}$. By diagonalization we can construct a subsequence $(g_k)_k$ of $(f_n)_n$ which converges on each $x \in X$. In particular, $(g_k)_k$ is $\sigma(E', E)$-Cauchy and therefore converges in $\sigma(E', E)$ to some $f \in E'$. Since $(f_n)_n$ is $\sigma(E', M)$-Cauchy, it also converges in $\sigma(E', M)$ to f, which proves our third assertion. ■

1.7.7 Proposition *Let M be a closed, countable-codimensional subspace of an lcs E. If E is ω-barrelled, has property (C) or has property (S), then M has the corresponding property.*

Proof. Let N be any algebraic complement of M in E and denote by P the projection of E onto M along N. Let $(f_n)_n$ be a $\sigma(M', M)$-Cauchy sequence in M'. By 1.7.1, $(f_n \circ P)_n$ is in E'. It is even a $\sigma(E', E)$-Cauchy sequence, so that by our assumption there exists some $g \in E'$ such that $f_n \circ P \to g$ in $\sigma(E', E)$. Thus $f_n \to g\mid_M$ in $\sigma(M', M)$ and M has property (S).
Now suppose E has property (C) and let $B \subseteq M'$ be $\sigma(M', M)$-bounded. For any sequence $(f_n)_n$ in B, $(f_n \circ P)_n$ is $\sigma(E', E)$-bounded and therefore has a cluster point g in $\sigma(E', E)$. But then $g\mid_M$ is a $\sigma(M', M)$-cluster point of $(f_n)_n$, so that M has property (C).
If, finally, E is ω-barrelled and X is a countable $\sigma(M', M)$-bounded subset of M', we let $X' := \{f \circ P \mid f \in X\}$. By 1.7.1, $X' \subseteq E'$ and clearly X' is $\sigma(E', E)$-bounded. Hence X' is equicontinuous in E'. It follows that X is equicontinuous in M' and the proof is complete. ■

By reason of 1.7.6 and 1.7.7 we have:

1.7.8 Theorem *Let M be a countable-codimensional subspace of an lcs E. If E is ω-barrelled, has property (C) or has property (S), then M has the corresponding property.* ■

By a similar argumentation one could obtain an independent proof of 1.5.3 (cf. [32],

p. 94).

Now we turn to a more detailed discussion of c_o-barrelledness.

1.7.9 Proposition *Let E be a c_o-barrelled space with property (S) and M a closed, countable-codimensional subspace of E. Then each algebraic complement N of M is a topological complement and E induces on N the strongest locally convex topology.*

Proof. Denote by P the projection of E onto N along M. As in the proof of 1.7.2 it is enough to show that for each absorbing absolutely convex subset V of N, $P^{-1}(V)$ is a neighborhood of 0 in E. If $dim(N) < \infty$, we are done.
Otherwise let $B := \{x_n \mid n \in \mathbf{N}\}$ be a Hamel basis in N with $B \subseteq V$. For $k \in \mathbf{N}$ define $f_k \in E^*$ such that $f_k \mid_M = 0$ and $f_k(x_n) = 2^k \delta_{kn}$. By 1.7.1, $f_k \in E'$ for each k. Furthermore $f_k \to 0$ in $\sigma(E', E)$, so that $(f_k)_k$ is equicontinuous. Let $x \in \{f_k \mid k \in \mathbf{N}\}^\circ$, $x = y + \sum_{i=1}^n \lambda_i x_i$ with $y \in M$, $\lambda_i \in \mathbf{K}$. Then $\mid 2^i \lambda_i \mid = \mid f_i(x) \mid \leq 1$ for $i \in \mathbf{N}$, so that $x \in M + V$ since V is absolutely convex. Therefore $\{f_k \mid k \in \mathbf{N}\}^\circ \subseteq M + V \subseteq P^{-1}(V)$ and $P^{-1}(V)$ is a neighborhood of 0 in E. ■

1.7.10 Corollary *Every closed, countable-codimensional subspace M of a c_o-barrelled space with property (S) is c_o-barrelled.*

Proof. Since each separated quotient of a c_o-barrelled space is c_o-barrelled (cf.1.8.3 below), this follows directly from 1.7.9 ■

In dealing with dense, countable-codimensional subspaces we first prove

1.7.11 Lemma *Let E be an lcs with property (S) and $(f_n)_n$ a $\sigma(E', E)$-bounded sequence in E'. If the subspace $F := \{x \mid f_n(x) \to 0\}$ is of countable codimension in E, then F is closed.*

Proof. Choose $x_1 \notin F$ (the case $E = F$ is trivial) and let $B = \{x_i \mid i \in J\}$ be a Hamel basis of an algebraic complement of F, where $J = \{1, 2, ..., k\}$ or $J = \mathbf{N}$ according as $codim(F) = k$ or $= \aleph_0$. $(f_n(x_1))_n$ is a bounded sequence in $\mathbf{K}$ not converging to 0, so that there exists $\alpha_1 \neq 0$ and a subsequence $(f_{n_1})_{n_1}$ of $(f_n)_n$ such that $f_{n_1}(x_1) \to \alpha_1$.
Continuing in this fashion, for each $j \geq 2$ we can find a subsequence $(f_{n_j})_{n_j}$ of $(f_{n_{j-1}})_{n_{j-1}}$ and $\alpha_j \in \mathbf{K}$ with $\lim_{n_j \to \infty} f_{n_j}(x_i) = \alpha_i$, $1 \leq i \leq j$. If $\mid B \mid = k$, then $(f_{n_k})_{n_k}(x_i)$ converges to α_i for $1 \leq i \leq k$. Otherwise, by diagonalization we obtain a subsequence $(g_p)_p$ of $(f_n)_n$ such that $g_p(x_i) \to \alpha_i$ for every $i \in \mathbf{N}$.

Now define $g \in E^*$ by $g \mid_F = 0$ and $g(x_i) = \alpha_i$ for $x_i \in B$. Then for each $x \in E$, $g(x) = \lim_{n_k \to \infty} f_{n_k}(x)$ if $\mid B \mid = k$ and $g(x) = \lim_{p \to \infty} g_p(x)$ otherwise.
In both cases, since E has property (S), $g \in E'$. Thus $\overline{F} \subseteq g^{-1}(0)$ and $x_1 \notin \overline{F}$ since $g(x_1) = \alpha_1 \neq 0$. It follows that $F = \overline{F}$. ■

1.7.12 Proposition *Every dense, countable-codimensional subspace M of a c_o-barrelled space E with property (S) is c_o-barrelled.*

Proof. Let $(f_n)_n$ be a $\sigma(E', M)$-null sequence in $M' = E'$. By 1.7.5, $(f_n)_n$ is $\sigma(E', E)$-bounded. Define F as in 1.7.11. Since $M \subseteq F$, the hypothesis of 1.7.11 applies so that F is closed. Therefore $F = E$ and $f_n \to 0$ in $\sigma(E', E)$. Consequently, $\{f_n \mid n \in \mathbf{N}\}$ is equicontinuous on E and hence on M. ■

1.7.13 Theorem *Every countable-codimensional subspace M of a c_o-barrelled space E with property (S) is c_o-barrelled (and also has property (S) by 1.7.8).*

Proof. 1.7.10 and 1.7.12 ■

If, in 1.7.13, we omit the assumption that E has property (S), the conclusion is no longer valid. In a certain respect it is then 'as wrong as can be':

1.7.14 Proposition *Every c_o-barrelled space E without property (S) contains a dense hyperplane which is not c_o-barrelled.*

Proof. By our assumption there exists a non-continuous linear functional f and a sequence $(f_n)_n$ in E' such that $f_n(x) \to f(x)$ for every $x \in E$. $H := f^{-1}(0)$ is a dense hyperplane in E which is not c_o-barrelled: $(f_n)_n \subseteq E' = H'$ is a $\sigma(E', H)$-null sequence but $(f_n \mid_H)_n$ is not equicontinuous. Otherwise $(f_n)_n$ would be equicontinuous since $\overline{H} = E$. This contradicts $f \notin E'$ (every pointwise limit of an equicontinuous sequence is continuous). ■

That spaces as described in the hypothesis of 1.7.14 really exist can be seen from 1.9.6.
The concepts introduced so far allow some statements concerning the inheritance properties of Mackey spaces. It is well known that each evaluable lcs is a Mackey space (see [23], p.218, Prop.8). Other aspects of the relationship between 'almost barrelled' and Mackey spaces are revealed by

1.7.15 Proposition *Let M be a countable-codimensional subspace of a Mackey space E. Then each of the following two conditions is sufficient for M to be a Mackey space:*

(i) M is closed and E has property (S).

(ii) M is dense and separable and E has property (C).

Proof. Suppose (i) is satisfied. Then by 1.7.2 M is isomorphic to a separated quotient of E and hence is Mackey (see [50], IV.4.1, Cor.4).
If (ii) holds, we identify M' with E'. Let B be any $\sigma(E', M)$-compact subset of E'. By 1.7.5, B is $\sigma(E', E)$-bounded and therefore $\sigma(E', E)$-precompact (cf.[50], IV.5.5, Cor.2). Let $(f_\alpha)_{\alpha \in D}$ be a $\sigma(E', E)$-Cauchy net in B. Since B is $\sigma(E', M)$-compact, there exists some $f \in B$ such that $(f_\alpha)_{\alpha \in D}$ is $\sigma(E', M)$-convergent to f. We claim that f is the $\sigma(E', E)$-limit of $(f_\alpha)_{\alpha \in D}$.
Otherwise there would exist $x_o \in E$ with $f_\alpha(x_o) \not\to f(x_o)$. Then for some $\epsilon > 0$ and arbitrarily high indices α we have $| f_\alpha(x_o) - f(x_o) | \geq \epsilon$. Now choose a sequence $(\beta_n)_n$ in D with $| f_{\beta'}(x_o) - f_{\beta''}(x_o) | < \frac{1}{n}$ for $\beta', \beta'' \geq \beta_n$ and choose an increasing sequence $(\gamma_n)_n$ such that $\gamma_n \geq \beta_n$ and $| f_{\gamma_n}(x_o) - f(x_o) | \geq \epsilon$ for each n.
Let $\{x_n \mid n \in \mathbf{N}\}$ be a countable dense subset of M. By diagonalization we can choose a subsequence $(\alpha_j)_{j \in \mathbf{N}}$ of $(\gamma_n)_n$ such that simultaneously $\lim_{j \to \infty} f_{\alpha_j}(x_n) = f(x_n)$ for $n \in \mathbf{N}$ and $| f_{\alpha_j}(x_o) - f(x_o) | \geq \epsilon$ for $j \in \mathbf{N}$ hold. Let $g \in E'$ be a cluster point of the $\sigma(E', E)$-bounded sequence $(f_{\alpha_j})_j$ (E has property (C)). Then $g(x_n) = f(x_n)$ for each n and $| g(x_o) - f(x_o) | \geq \epsilon$.
$g - f$ is a continuous linear functional vanishing on the dense subset $\{x_n \mid n \in \mathbf{N}\}$ of E. This implies that $f = g$. But $| g(x_o) - f(x_o) | \geq \epsilon$, a contradiction. This means that the $\sigma(E', M)$-limit of every $\sigma(E', E)$-Cauchy net in B is also its $\sigma(E', E)$-limit. Hence B is complete and precompact, i.e. B is $\sigma(E', E)$-compact. Hence B is equicontinuous on E and therefore on M, so that M is a Mackey space. ∎

1.7.16 Corollary *Let E be a Mackey space with property (C). Then every separable subspace of countable codimension in E is itself a Mackey space.*

Proof. If the subspace M is closed or dense, we are done by 1.7.15. If M is neither closed nore dense, apply 1.7.15 (i) to $\overline{M}$ as a subspace of E and then 1.7.15 (ii) to M as a subspace of $\overline{M}$. ∎

1.8 Inheritance Properties

The following lemma will make it possible to generalize some well known inheritance properties of barrelled spaces to most of the spaces considered in this chapter.

1.8.1 Lemma *Let E, F be lcs's and $f : E \to F$ a continuous linear map. If B is a (bornivorous) barrel, infrabarrel, ω-barrel or c_o-barrel in F, respectively then $f^{-1}(B)$ is a subset of the same type in E.*

Proof. In either case, $f^{-1}(B)$ is closed and absolutely convex. Let A be bounded in E or $A = \{x\}$ according as B is bornivorous or just absorbing. Then for some $\lambda > 0$, $f(A) \subseteq \lambda B$ so that $A \subseteq \lambda f^{-1}(B)$, implying that $f^{-1}(B)$ is bornivorous or absorbing.
If $B = \bigcap_{n=1}^{\infty} U_n$ is an infrabarrel, where each U_n is a closed absolutely convex neighborhood of 0 in F, then $f^{-1}(B) = \bigcap_{n=1}^{\infty} f^{-1}(U_n)$ is an infrabarrel in E.
Finally, let $B = \{g_n \mid n \in \mathbf{N}\}^\circ$ be an ω-barrel (resp. c_o-barrel) in F, where $(g_n)_n$ is $\sigma(E', E)$-bounded (resp. a $\sigma(E', E)$-null sequence). Then $f^{-1}(B) = \{g_n \circ f \mid n \in \mathbf{N}\}^\circ$ is of the same type in E. ■

1.8.2 Theorem *Let F be a vector space, $(E_\iota)_{\iota \in I}$ a family of lcs's and for each $\iota \in I$ let f_ι be a linear map from E_ι into F. Equip F with the finest locally convex topology τ such that all the maps f_ι are continuous and suppose that (F, τ) is Hausdorff.*
If all the E_ι are barrelled, infrabarrelled, ω-barrelled, c_o-barrelled, evaluable, infra-evaluable, ω-evaluable or c_o-evaluable, respectively then F has the same property.

Proof. Let B be the respective kind of barrel in F. Then by 1.8.1 and our hypothesis, $f_\iota^{-1}(B)$ is a neighborhood of 0 in E_ι for each $\iota \in I$. Hence B is a neighborhood of 0 in F. ■

1.8.3 Corollary *The classes of spaces mentioned in 1.8.2 are stable under the formation of separated quotients, locally convex direct sums and Hausdorff inductive limits.* ■

1.9 Examples

At the end of this chapter we present a selection of examples in order to demonstrate that some of the concepts introduced so far actually describe different classes of spaces. Other distinguishing examples can be found in [24], chapter 12.

1.9.1 *A Mackey space with property (S) but not property (C).*
Set $(E, \tau) := (l^\infty, \mu(l^\infty, l^1))$.
By [15], p.374 l^1 is $\sigma(l^1, l^\infty)$-sequentially complete, so that (E, τ) has property (S). By its definition, E is a Mackey space. Denote by e_n the n-th unit vector in l^1, i.e.

$e_n = (\delta_{nm})_{m\in\mathbf{N}}$. Then $B := \{e_n \mid n \in \mathbf{N}\}$ is $\sigma(l^1, l^\infty)$-bounded, but no $y \in l^1$ can be a cluster point of $(e_n)_n$: For such a y, we would have $y_m = \langle y, f_m\rangle = 0$ for each m, since - for m fixed - $\langle e_n, f_m\rangle = 0$ finally (here, f_m denotes the n-th unit vector, considered as an element of l^∞). For $e = (1,1,1,...)$, however, $\langle e_n, e\rangle \equiv 1$ converges to 1 and not to $\langle y, e\rangle = 0$. Thus E does not possess property (C).

1.9.2 *An ω-barrelled Mackey space which is not barrelled.*

Let E be a vector space with uncountable algebraic basis B. E can canonically be identified with $\mathbf{K}^{(B)} = \bigoplus_{b\in B}\mathbf{K}$. Let F be the subspace of $E^* = \prod_{b\in B}\mathbf{K}$ consisting of those $f \in E^*$ for which $f(x) \neq 0$ for at most countably many $x \in B$.

Then $\langle E, F\rangle$ is a dual system. Denote by τ_o the topology on E of uniform convergence on the $\sigma(F,E)$-bounded sequences in F, i.e. $\tau_o = \nu(E,F)$ (cf.1.6.4). We assert that τ_o is compatible with the duality $\langle E, F\rangle$.

Clearly, $\sigma(E,F) \leq \tau_o$. Now let $f_o \in E^*$ be τ_o-continuous. Then for some countable, $\sigma(F,E)$-bounded subset A of F we have $\mid f_o(e) \mid \leq 1$ for every $e \in A^\circ$. Let $B_1 := \{x \in B \mid f(x) \neq 0 \text{ for some } f \in A\}$. B_1 is countable since A is a countable subset of F. If $x \in B \setminus B_1$, $\lambda x \in A^\circ$ for all scalars λ. Therefore $\mid f_o(\lambda x) \mid \leq 1$ for every $\lambda \in \mathbf{K}$, implying $f_o(x) = 0$. Hence $f_o \in F$ so that $\tau_o \leq \mu(E,F)$.

By 1.6.6, (E, τ_o) and $(E, \mu(E,F))$ are ω-barrelled. In particular, $(E, \mu(E,F))$ is an ω-barrelled Mackey space.

It remains to show that it is not barrelled: Let $\mathcal{F}$ be the family of all finite subsets of B, directed by inclusion. For each $C \in \mathcal{F}$ choose $f_C \in F$ with $f_C(x) = 1$ for $x \in C$ and $f_C(x) = 0$ for $x \in B \setminus C$. Then $\mathcal{N} := \{f_C \mid C \in \mathcal{F}\}$ is a net which is $\sigma(E^*, E)$-convergent to that $g \in E^*$ satisfying $g(x) = 1$ for all $x \in B$. Now $g \notin F$ since B is uncountable, i.e. g is not $\mu(E,F)$-continuous. But then $\mathcal{N}$ cannot be equicontinuous. On the other hand, $\mathcal{N}$ is obviously $\sigma(F,E)$-bounded, so that $(E, \mu(E,F))$ is not barrelled.

1.9.3 *An ω-barrelled space which is not a Mackey space*

Let (E, τ_o) be the ω-barrelled space defined in 1.9.2. We show that (E, τ_o) is not a Mackey space.

$V := \{y = \sum_{x\in B}\lambda_x x \mid\mid \lambda_x \mid\leq 1 \forall x \in B\}$ is a barrel in E. Denote by p_V the gauge of V and by τ_1 the coarsest locally convex topology finer than τ_o and the one defined by p_V. Then $\{\epsilon V \cap W \mid \epsilon > 0, W \text{ is a } \tau_o\text{-neighborhood of } 0\}$ is a fundamental system of neighborhoods of 0 for τ_1. Clearly $\tau_1 \geq \tau_o$ so that $(E, \tau_1)' \supseteq (E, \tau_o)' = F$.

Moreover, $\tau_1 \neq \tau_o$: Let A be any $\sigma(F,E)$-bounded sequence and set $B_o := \{x \in B \mid \exists f \in A \text{ with } f(x) \neq 0\}$. Since B_o is countable there exists some $x \in B \setminus B_o$. For this x we have $\{\lambda x \mid \lambda \in \mathbf{K}\} \subseteq A^\circ$. Thus V is not a τ_o-neighborhood of 0 because it doesn't contain any one-dimensional subspace of E.

Next we show that $(E,\tau_1)' \subseteq F$: Let U be any absolutely convex τ_o-neighborhood of 0 and choose a $\sigma(F,E)$-bounded sequence A with $A^\circ \subseteq U$. With B_o defined as above it follows that $\lambda x \in U$ for all $\lambda \in \mathbf{K}$ and all $x \in B \setminus B_o$.

Suppose that $f \in E^* \setminus F$. Then there exist uncountably many elements x of $B \setminus B_o$ with $f(x) \neq 0$. Hence there is some $\epsilon > 0$ such that $\mid f(x) \mid \geq \epsilon$ for infinitely many $x \in B \setminus B_o$. Choose $S := \{x_n \mid n \in \mathbf{N}\} \subseteq B \setminus B_o$ such that $\mid f(x_n) \mid \geq \epsilon$ for each $n \in \mathbf{N}$. Since U is absolutely convex and $S \subseteq B \setminus B_o$ it follows that $y_N := \sum_{n=1}^{N} x_n \frac{|f(x_n)|}{f(x_n)} \in U \cap V$ for every $N \in \mathbf{N}$.

$f(y_N) \geq N\epsilon$ now yields that f is unbounded on $U \cap V$. Since U was arbitrary, we conclude that f is discontinuous for τ_1. Therefore $(E,\tau_1)' = F$ and $\tau_1 \leq \mu(E,F)$.

Hence (E,τ_o) is not a Mackey space, because $\tau_o < \tau_1 \leq \mu(E,F)$.

Confer 6.2.4 for a similar example.

1.9.4 *An lcs with property (C) that is not c_o-barrelled*

Let E be any infinite-dimensional vector space and consider $(E,\sigma(E,E^*))$. If B is any Hamel basis of E, then $E = \bigoplus_{b\in B} \mathbf{K}$ and $E^* = \prod_{b\in B} \mathbf{K}$. Every $\sigma(E^*,E)$-bounded subset of E^* is bounded in each coordinate and therefore it is relatively $\sigma(E^*,E)$-compact by TYCHONOFF'S theorem. Hence E has property (C). Let $B_o = \{x_n \mid n \in \mathbf{N}\}$ be a countably infinite subset of B. For $n \in \mathbf{N}$ let f_n be the element of E^* with $f_n(x_n) = 1$ and $f_n(y) = 0$ for $y \in B \setminus \{x_n\}$. $(f_n)_n$ is a $\sigma(E^*,E)$-null sequence but $\{f_n \mid n \in \mathbf{N}\}$ is not equicontinuous because it is not contained in any finite-dimensional subspace of E^*. Thus E is not c_o-barrelled.

1.9.5 *$(l^1,\mu(l^1,c_o))$ is a Mackey space without property (S).*

To see this, consider the sequence $x_n = (\underbrace{1,1,\ldots,1}_{\text{n times}},0,0,\ldots)$ in c_o, whose $\sigma(l^\infty,l^1)$-limit $x = (1,1,1,\ldots)$ does not belong to c_o.

1.9.6 *$(l^1,\nu_o(l^1,c_o))$ is a c_o-barrelled space without property (S).*

This follows from 1.6.6 and from 1.9.5, since $\nu_o(l^1,c_o)$ is compatible with $\langle l^1,c_o\rangle$ (see 1.6.7).

By 1.7.14, $(l^1,\nu_o(l^1,c_o))$ contains a dense, one-codimensional subspace which is not c_o-barrelled.

1.9.7 $(l^\infty, \mu(l^\infty, l^1))$ *is a complete c_o-barrelled space with property (S) which is not ω-barrelled.*

E is c_o-barrelled by 1.6.8. By [24], p.206, corollaries 10.5.3 and 10.5.4, E is complete and possesses property (S) (also cf.1.9.1). Since E does not have property (C) by 1.9.1, E is not ω-barrelled.

1.10 Survey

The following diagram shows the interdependency of the classes of spaces treated in this chapter.

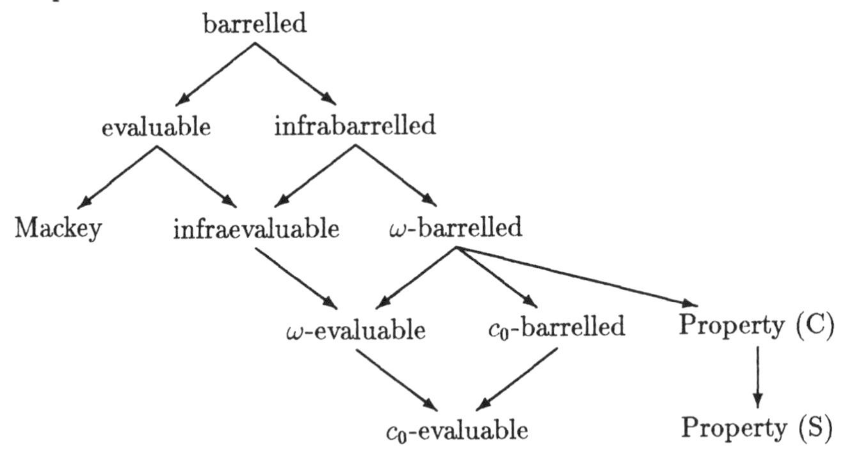

1.11 References

1.1.1 is a classical result by J.DIEUDONNÉ ([14], Th.3). The remarkably simple proof given here is due to M.DE WILDE ([9]) who, together with C.HOUET, also introduced the concepts of infrabarrelledness and absorbent sequences in the sense we use it here ([10]). These two authors use the term σ-barrelled instead of ω-barrelled. H.JARCHOW uses a modified definition of absorbent sequences by additionally requiring $A_n + A_n \subseteq A_{n+1}$ for each $n \in \mathbf{N}$ ([24], p.253).

It should generally be said that the notation of 'almost barrelled' spaces in the literature is far from uniform: 'Infrabarrelled' often denotes what we call 'evaluable' (cf.[23]). A.GROTHENDIECK ([21]) and H.JARCHOW ([24]) use the term 'quasi-barrelled' instead of 'evaluable'. In [24], ω-barrelled spaces are called 'l^∞-barrelled' and infrabarrelled spaces are termed '$\aleph_0$-barrelled'.

With a view to not further complicating the notations in this matter we have therefore refrained from calling the spaces treated in the first chapter 'prebarrelled spaces' (and from terming the spaces in chapters 5 and 6 'pre-Baire') though it cost us quite some effort to do so.
1.4.2 (ii)-1.4.7 and 1.4.11-1.4.15 are taken from [10]. For metrizable barrelled spaces (whose completion is Fréchet and hence Baire), 1.4.15 was first given by I.AMEMIYA and Y.KOMURA (cf.[3], Satz 1).
1.4.3, 1.4.4, 1.4.10, 1.4.17 and 1.4.19 are essentially due to M.VALDIVIA ([53]) and were given their present form by M.DE WILDE and C.HOUET in [10].
1.5.1-1.5.3 and 1.5.7-1.5.9 go back to [53]. An independent proof of 1.5.3 was given by M.LEVIN and S.A.SAXON in [32]. The other results in 1.5 are taken from [10].
1.6.7 and 1.6.8 were established by J.M.GARCIA-LAFUENTE in [19]. 1.7.1-1.7.8, 1.7.15 and 1.7.16 were obtained by M.LEVIN and S.A.SAXON in [32] and [33], while 1.7.9-1.7.14 are taken from [19]. Finally, the examples presented in 1.9 occured in [32], [33] and [19].
For further information on the history of the development of weakened barrelledness concepts see [24], 12.7.

2 The Strongest Locally Convex Topology

For each infinite cardinal m, let Φ_m denote the m-dimensional vector space $\mathbf{K}^{(m)}$ endowed with the strongest locally convex topology (cf. [50], p.56). If, in particular, $m = \aleph_0$ we simply write Φ instead of Φ_m.
In many respects, Φ behaves like a finite-dimensional tvs and thus appears as an appropriate generalization of such spaces. We say that an lcs E *contains* Φ if some subspace F of E is isomorphic to Φ and that E *contains* Φ *complemented* if, in addition to this, F possesses a topologically complementary subspace in E.
In subsequent chapters, Φ will prove to play an important role in characterizing spaces that are 'almost Baire' as well as in classifying (LF)-spaces. We will see that many non-metrizable nuclear spaces contain Φ, that many (LF)-spaces contain Φ complemented and that the spaces Φ_m are closely connected with questions arising in the context of varieties of lcs's.
At first sight, these results are far from obvious or, as S.A.SAXON puts it (cf. [45]): 'This prominence of Φ appears unexpected. The general feeling, expressed by EDWARDS ([16], p.78), has been that the situation in which E is a vector space with the strongest locally convex topology '... seldom if ever occurs in the study of problems having their roots in analysis, save when E is of finite dimension...'. '
It is for the reasons stated that in the first part of this chapter we will concern ourselves with some basic properties of the space Φ. Though the proofs will be given explicitly only for the space Φ itself, many results carry over to Φ_m, m an arbitrary infinite cardinal.

2.1 Fundamental Results

Algebraically, Φ is isomorphic to $\mathbf{K}^{(\mathbf{N})} = \bigoplus_{n=1}^{\infty} \mathbf{K}$ and we will identify Φ with $\mathbf{K}^{(\mathbf{N})}$. Let τ denote the strongest locally convex topology on Φ. Then we have

- *τ is identical with τ_1, the topology of the locally convex direct sum* $\mathbf{K}^{(\mathbf{N})}$.

  Proof. τ_1 is the strongest locally convex topology on $\Phi = \mathbf{K}^{(\mathbf{N})}$ such that the canonical injections $j_n : \mathbf{K} \hookrightarrow \mathbf{K}^{(\mathbf{N})}$ are continuous. But since $dim(\mathbf{K}) = 1$, the j_n are continuous for *every* locally convex topology on $\mathbf{K}^{(\mathbf{N})}$, implying that $\tau_1 = \tau$. ■

Thus we immediately have (cf. [50], II.6.1, II.6.2):

- *(Φ, τ) is an lcs (i.e. Hausdorff).*

- *A neighborhood base of 0 in (Φ, τ) is formed by all sets*

$$U = \Gamma_{n=1}^{\infty} j_n(U_n) = \left\{ \sum_n \lambda_n j_n(x_n) \mid \sum_n \mid \lambda_n \mid \leq 1, x_n \in U_n \right\}$$

  where U_n is a neighborhood of 0 in $\mathbf{K}$ for each $n \in \mathbf{N}$.

- *(Φ, τ) is complete.*

- *Let F be any lcs, $u : (\Phi, \tau) \to F$ a linear map. Then u is continuous.*

  Proof. For each n, $u \circ j_n$ is continuous, hence u is continuous by the universal property of the locally convex direct sum. ■

In particular:

- $\Phi' = \Phi^*$

- *The system of all semi-norms on Φ forms a generating family for τ. Equivalently, all absorbing absolutely convex subsets of Φ form a neighborhood base of 0 in (Φ, τ).*

- *Each subspace of Φ is closed and topologically complemented.*

  Proof. Let M be a subspace of Φ and N an algebraic complement of M in Φ. The projection P of Φ onto N along M is continuous because it is linear. Therefore $M = P^{-1}(0)$ is closed and N is a topological complement. ■

- *(Φ, τ) is a strict (LB)-space (and therefore a strict (LF)-space, cf. 7.1.2).*

  Proof. Denote by E_n the space $\mathbf{K}^n$ with its usual topology, considered as a subspace of $\mathbf{K}^{(\mathbf{N})}$. Then $\Phi = \bigcup_{n=1}^{\infty} E_n$ and $E_n \subset E_{n+1}$ (continuously) for each n. Set $(\Phi, \sigma) = \varinjlim E_n$. Then $\tau \leq \sigma$ since $E_n \hookrightarrow \Phi$ is continuous for each n $(dim(E_n) < \infty)$. $\sigma \leq \tau$ is clear. ■

Consequently:

- Φ *is barrelled.*([50], II.7.2, Cor.2)
- Φ *is bornological.*([50], II.8.2, Cor.2)
- Φ *is reflexive.*([50], IV.5.8)
- Φ *is a Mackey space.*([50], IV.3.4)
- Φ *is a Schwartz space.*([23], p.280, Prop.8)
- Φ *is nuclear.*([50], III.7.4)

On the negative side we have:

- Φ *is not metrizable.*

Proof. Otherwise Φ would be a Fréchet space and therefore Baire. With E_n defined as above there would exist some $n \in \mathbf{N}$ such that $int(E_n) \neq \emptyset$, implying that E_n is a neighborhood of 0. Consequently, $\Phi = \bigcup_{m=1}^{\infty} mE_n = E_n$, a contradiction. ∎

Actually, this proof shows that a Fréchet space (more generally, a linear Baire space) cannot have Hamel dimension $\aleph_0$.

- Φ *is separable.*

Proof. $\bigoplus_{n=1}^{\infty} \mathbf{Q}$ (resp. $\bigoplus_{n=1}^{\infty}(\mathbf{Q} + i\mathbf{Q})$) is dense in Φ. ∎

- *For each* $n \in \mathbf{N}$, $\Phi \cong \prod_{i=1}^{n} \Phi$.

Proof. It suffices to show that $\Phi \cong \Phi \times \Phi$. Since this is clear algebraically, it remains to prove that $\tau \times \tau$ is the finest locally convex topology on $\Phi \times \Phi$. For finite indexing sets the product topology is identical with the locally convex direct sum topology (see 7.1.20 (iii) for a proof of this fact).
Therefore $\tau \times \tau$ is the finest locally convex topology on $\Phi \times \Phi$ such that $j_1 : \Phi \hookrightarrow \Phi \times \Phi$ and $j_2 : \Phi \hookrightarrow \Phi \times \Phi$ are continuous. But by what we already know these mappings are continuous for every locally convex topology on $\Phi \times \Phi$ which yields the result. ∎

- *A subset $B \subseteq \Phi$ is bounded if and only if it is contained in some finite-dimensional subspace F of Φ and bounded there.*

Proof. The condition is evidently sufficient. The necessity follows from the corresponding theorem on locally convex direct sums (cf. [50], II.6.3). ■

In particular:

- *$B \subseteq \Phi$ is bounded if and only if it is relatively compact.*

Hence

- *Φ is a Montel space.*

- *$B \subseteq \Phi$ is sequentially closed if and only if $B \cap M$ is closed for each finite-dimensional subspace M of Φ.*

Proof. Since each finite-dimensional subspace is isomorphic to some $\mathbf{K}^n$, the condition is necessary.
Conversely, let $(x_n)_n$ be a sequence in B converging to x. Then $(x_n)_n$ is bounded and therefore contained in some finite-dimensional M. Thus $x \in M \cap B = B$. ■

Denote by ω the Fréchet space $\prod_{n=1}^{\infty} \mathbf{K}$ with the usual product topology. Algebraically, ω and Φ are topological duals of each other, i.e. $\omega' = \Phi$ and $\Phi'(= \Phi^*) = \omega$.

Moreover

- *Φ is the strong dual of ω, $(\Phi, \tau) = (\omega', \beta(\omega', \omega))$.*

Proof. Since Φ is Mackey, we have $\tau = \mu(\Phi, \Phi') = \mu(\omega', \omega) \leq \beta(\omega', \omega) \leq \tau$ (τ is the strongest locally convex topology on Φ). ■

- *ω is the strong dual of Φ, $\omega = (\Phi', \beta(\Phi', \Phi))$.*

Proof. This follows directly from the above description of the bounded subsets of Φ (cf. [23], p.197). ■

- *The property of carrying the strongest locally convex topology is inherited by subspaces, quotients and inductive limits.*

Proof. Let (E, τ) be an lcs with the strongest locally convex topology and M a subspace of E. Let V be an absorbing absolutely convex subset of M and choose a Hamel basis $\{x_\alpha \mid \alpha \in A\}$ in an algebraic complement N of M. Then V_1, the absolutely convex hull of $V \cup \{x_\alpha \mid \alpha \in A\}$ is absorbing and is therefore a neighborhood of 0 in E. Hence $V = V_1 \cap M$ is a neighborhood of 0 in M.

Furthermore, $M \cong E/N$ so that E/N, too, carries the strongest locally convex topology.

Finally, let $(E, \tau) = \underrightarrow{\lim}(E_\alpha, h_{\beta\alpha})$ (E_α carrying the strongest locally convex topology) and denote by $g_\alpha : E_\alpha \hookrightarrow E$ the canonical injections. A fundamental system of neighborhoods of 0 for τ is given by $\mathcal{U} = \{V \subseteq E \mid V$ is absorbing and absolutely convex and $g_\alpha^{-1}(V)$ is a neighborhood of 0 in E_α for each $\alpha\}$ $= \{V \subseteq E \mid$ V is absorbing and absolutely convex$\}$. ■

2.2 Interrelation with Barrelled Spaces

To begin with, let us reformulate 1.7.2 and 1.7.9 in the context of this chapter.

2.2.1 Theorem *Let E be a Mackey space with property (S) or a c_o-barrelled space with property (S). Then E has a closed, $\aleph_0$-codimensional subspace if and only if E contains Φ complemented.* ■

In particular, the hypothesis of 2.2.1 is satisfied if E is barrelled or, more generally, ω-barrelled.

The following is a characterization of the space Φ by means of the concepts we studied in the first chapter.

2.2.2 Theorem *Up to isomorphism, Φ is the only $\aleph_0$-dimensional lcs which is either*

(i) barrelled

(ii) infrabarrelled

(iii) ω-barrelled

(iv) c_o-barrelled with property (S), or

(v) Mackey with property (S)

Proof. By 1.10 we have (i)$\Rightarrow$ (ii)$\Rightarrow$(iii)$\Rightarrow$(iv) and (i)$\Rightarrow$(v). Thus, since Φ is barrelled it has all properties (i)-(v).
Conversely, assume that E is an $\aleph_0$-dimensional lcs satisfying (iv) or (v). Then $\{0\}$ is a closed, $\aleph_0$-codimensional subspace of E. Hence by 2.2.1 E is isomorphic to Φ. ■

2.3 Φ-Productive Spaces

2.3.1 Definition *An lcs E is* Φ-productive *if there exists a sequence $(x_i)_{i\in\mathbf{N}}$ in E and a continuous seminorm p on $sp((x_i)_i)$ such that $p(x_i) > 0$ for each $i \in \mathbf{N}$ and $p(\sum_{i=1}^n \lambda_i x_i) = \sum_{i=1}^n \mid \lambda_i \mid p(x_i)$ for arbitrary finite sequences $\lambda_1, ..., \lambda_n$.*

Note that for such $(x_i)_i$, $\{(x_i) \mid i \in \mathbf{N}\}$ is necessarily linearly independent and that by defining $y_i := \frac{x_i}{p(x_i)}$ we can additionally achieve $p(\sum_{i=1}^n \lambda_i y_i) = \sum_{i=1}^n \mid \lambda_i \mid$.

2.3.2 Lemma *Let $\widetilde{p}$ be a continuous seminorm on a subspace M of the lcs E. Then there exists a continuous seminorm p on E with $p\mid_M = \widetilde{p}$.*

Proof. For some absolutely convex neighborhood V of 0 in M, $\widetilde{p} = p_V$, the gauge of V. By [50], II.6.4 there exists an absolutely convex neighborhood U of 0 in E such that $V = U \cap M$. Now set $p := p_U$. ■

2.3.3 Theorem *For an arbitrary lcs E, the following statements are equivalent:*

(1) *E does not have the weak topology $\sigma(E, E')$.*

(2) *Some neighborhood of 0 does not contain a finite-codimensional subspace of E.*

(3) *There exists a continuous seminorm p on E such that $p\mid_F \neq 0$ for each finite-codimensional subspace F of E.*

(4) *E is not isomorphic to a subspace of a product of copies of* **K**.

(5) *E is Φ-productive.*

(6) *The product space E^I is Φ-productive for each indexing set $I \neq \emptyset$.*

(6′) *E^I is Φ-productive for some indexing set $I \neq \emptyset$.*

(7) *E^I contains Φ for each indexing set I with $\mid I \mid \geq 2^{\aleph_0}$.*

(7′) *E^I contains Φ for some indexing set I with $\mid I \mid \geq 2^{\aleph_0}$.*

Proof.

(1) $\Rightarrow$ (2) Suppose (2) is violated and let U be a closed absolutely convex neighborhood of 0. Then U contains some closed subspace F of codimension $n < \infty$. Let $N = sp(x_1, ..., x_n)$ be an algebraic and hence topological complement of F. Define $f_i \in E^*$ by $f_i(x_k) = \delta_{ik}$, $f_i \mid_F = 0$. Then $f_i \in E'$ $(1 \leq i \leq n)$. There exists $\lambda > 0$ such that $\lambda x_i \in U$ for each $i \leq n$. It follows that $\frac{\lambda}{n}\{f_1, ..., f_n\}^\circ \subseteq 2U$. Therefore U is a $\sigma(E, E')$-neighborhood of 0 and E carries its weak topology.

(2) $\Rightarrow$ (1) Each $\sigma(E, E')$-neighborhood of 0 contains a finite-codimensional subspace of the form $\{f_1, ..., f_n\}^\circ$.

(2) $\Leftrightarrow$ (3) This is clear due to the form of the 1-1 correspondence between absolutely convex 0-neighborhoods and continuous seminorms.

(2) $\Rightarrow$ (4) Let E be a subspace of $\mathbf{K}^I$ and U a 0-neighborhood in E. There exists a finite subset $H \subseteq I$ and for $i \in H$ a 0-neighborhood V_i in $\mathbf{K}$ such that

$$U \supseteq \left(\prod_{i \in H} V_i \times \prod_{i \in I \setminus H} \mathbf{K} \right) \cap E \supseteq \left(\{0\} \times \prod_{i \in I \setminus H} \mathbf{K} \right) \cap E =: F.$$

Obviously, $codim_E(F) \leq \mid H \mid$.

(4) $\Rightarrow$ (1) Assume that E has its weak topology. Since E is Hausdorff, the map $x \rightarrow (x' \rightarrow x'(x))$ is an injection from E into $(E')^*$. Furthermore, $\sigma((E')^*, E') \mid_E = \sigma(E, E')$, so that E can be considered as a (dense) subspace of $((E')^*, \sigma((E')^*, E'))$. It is well known that this lastnamed space is isomorphic to $\mathbf{K}^I$, where $\mid I \mid$ is the Hamel dimension of E' (cf. [23], p.188).

(3) $\Rightarrow$ (5) Choose $x_1 \in E$ such that $p(x_1) = \epsilon_1 > 0$. By the Hahn-Banach theorem there exists some $g_1 \in E'$ with $g_1(x_1) = \epsilon_1$ and $\mid g_1(x) \mid \leq p(x)$ for all $x \in E$. Suppose that $x_1, ..., x_k \in E$ and $g_1, ..., g_k \in E'$ have been chosen so that $g_i(x_j) = \epsilon_i \delta_{ij}$ $(i, j = 1, ..., k)$ and $\mid g_i(x) \mid \leq p(x)$ for all $x \in E$, where

$$\epsilon_i = \inf \left\{ p\left(\sum_{j=1}^{i-1} \lambda_j x_j + x_i \right) \mid \lambda_1, ..., \lambda_{i-1} \in \mathbf{K} \right\} > 0 \text{ for } i = 1, ..., k.$$

$H := p^{-1}(0)$ is a closed, infinite-codimensional subspace and therefore $H_k = sp(H \cup \{x_1, ..., x_k\})$ is of the same type. Hence there exists an element x_{k+1} in the k-codimensional subspace $\bigcap_{j=1}^k g_j^{-1}(0)$ which is not in H_k. By the induction hypothesis,

$$p\left(\sum_{j=1}^{k}\lambda_j x_j + x_{k+1}\right) \geq \mid g_l\left(\sum_{j=1}^{k}\lambda_j x_j + x_{k+1}\right)\mid = \mid \lambda_l \mid \epsilon_l \text{ for } 1 \leq l \leq k,$$

i.e.

$$p\left(\sum_{j=1}^{k}\lambda_j x_j + x_{k+1}\right) \geq \max_{1\leq l\leq k} \mid \lambda_l \mid \epsilon_l.$$

Set

$$\epsilon_{k+1} := \inf\left\{p\left(\sum_{j=1}^{k}\lambda_j x_j + x_{k+1}\right) \mid \lambda_1, ..., \lambda_k \in \mathbf{K}\right\}.$$

Clearly $0 \leq \epsilon_{k+1} \leq p(x_{k+1}) < \infty$. Choose N such that $(\min_{1\leq l\leq k}\epsilon_l)N > \epsilon_{k+1}$.By the Heine-Borel theorem,

$$r : (\lambda_1, ..., \lambda_k) \to p\left(\sum_{j=1}^{k}\lambda_j x_j + x_{k+1}\right)$$

attains its minimum μ on $\{\lambda \mid\mid \lambda \mid\leq N\}^k$. Off this set, r takes only values greater than $N(\max_{1\leq l\leq k}\epsilon_l) > \epsilon_{k+1}$. Thus, by definition of ϵ_{k+1}, $\mu = \epsilon_{k+1}$ is the global minimum of r over $\mathbf{K}^k$. Moreover, $\epsilon_{k+1} > 0$ since $p\left(\sum_{j=1}^{k}\lambda_j x_j + x_{k+1}\right) = 0$ would imply $x_{k+1} \in H_{k+1}$.
On $sp(x_1, ..., x_{k+1})$ we define a linear functional $\hat{g}_{k+1}$ by $\hat{g}_{k+1}(x_i) = \epsilon_{k+1}\delta_{k+1,i}$ for $1 \leq i \leq k+1$. Then for $\mu \neq 0$ we have

$$\mid \hat{g}_{k+1}\left(\sum_{j=1}^{k}\lambda_j x_j + \mu x_{k+1}\right)\mid = \mid \mu \mid \epsilon_{k+1} \leq \mid \mu \mid p\left(\sum_{j=1}^{k}\frac{\lambda_j}{\mu}x_j + x_{k+1}\right)$$

$$= p\left(\sum_{j=1}^{k}\lambda_j x_j + \mu x_{k+1}\right).$$

For $\mu = 0$,

$$\mid \hat{g}_{k+1}\left(\sum_{j=1}^{k}\lambda_j x_j\right)\mid = 0 \leq p\left(\sum_{j=1}^{k}\lambda_j x_j\right).$$

Again by the Hahn-Banach theorem there exists $g_{k+1} \in E'$ with $g_{k+1}(x_j) = \epsilon_{k+1}\delta_{k+1,j}$ $(1 \leq j \leq k+1)$ and $\mid g_{k+1}(x) \mid\leq p(x)$ for all $x \in E$.
Thus we have proved the existence of sequences $(x_i)_i$ in E, $(g_i)_i$ in E' and $(\epsilon_i)_i$ in $\mathbf{K}$ such that $\epsilon_i > 0$, $g_i(x_j) = \epsilon_i\delta_{ij}$ and $\mid g_i(x) \mid\leq p(x)$ for all $x \in E$. Set $y_j = \frac{2^j}{\epsilon_j}x_j$, so

that $g_i(y_j) = 2^i\delta_{ij}$ for all $i, j \in \mathbf{N}$.
$(g_k)_k$ is equicontinuous because it is dominated by the continuous seminorm p. Therefore $V := \{g_k \mid k \in \mathbf{N}\}^\circ$ is a neighborhood of 0. The sequence $(y_i)_i$ is linearly independent and we define the seminorm $q : \sum_{j=1}^n \lambda_j y_j \to \sum_{j=1}^n \mid \lambda_j \mid$ on $S = sp(\{y_i \mid i \in \mathbf{N}\})$.
For $k = 1, ..., n$, $\mid g_k(\sum_{j=1}^n \lambda_j y_j) \mid = \mid 2^k \lambda_k \mid \leq 1$ implies $\mid \lambda_k \mid \leq 2^{-k}$, i.e. q is bounded by 1 on $V \cap S$. Consequently, q is continuous on S and E is Φ-productive.
(5) $\Rightarrow$ (3) Suppose E doesn't have property (3). Since E is Φ-productive there exists a linearly independent sequence $(x_n)_n$ in E and a continuous seminorm $\widetilde{p}$ on $S = sp(\{x_n \mid n \in \mathbf{N}\})$ such that $\widetilde{p}(x_n) > 0$ and

$$\widetilde{p}\left(\sum_{i=1}^n \lambda_i x_i\right) = \sum_{i=1}^n \mid \lambda_i \mid \widetilde{p}(x_i)$$

for each finite sequence $\lambda_1, ..., \lambda_n$. Let p be a continuous seminorm on E with $p\mid_S = \widetilde{p}$ (cf.2.3.2). Then $p\mid_F = 0$ for some finite-codimensional subspace F of E. For $n > codim(F)$, $sp(x_1, ..., x_n) \cap F$ contains an element $y = \sum_{k=1}^n \lambda_k x_k \neq 0$. But this yields $0 = p(y) = \sum_{k=1}^n \mid \lambda_k \mid \widetilde{p}(x_k) > 0$, a contradiction.
(5) $\Rightarrow$ (7) Let I be an arbitrary indexing set and for $\iota \in I$ let E_ι be an lcs.
Under the assumption that the cardinal of $I_\Phi := \{\iota \in I \mid E_\iota \text{ is } \Phi\text{-productive}\}$ is at least $2^{\aleph_0}$ we will show that $F = \prod_{\iota \in I} E_\iota$ contains Φ. Since, for each $J \subseteq I$, F contains a subspace isomorphic to $\prod_{\iota \in J} E_\iota$ we may suppose without loss of generality that $I = I_\Phi$ and $\mid I \mid = 2^{\aleph_0}$. In this case there is a 1-1 correspondence between I and the set of all subsequences of $\mathbf{N}$ which we denote by $\iota \leftrightarrow (\lambda_k^\iota)_{k=1}^\infty$, where $\iota \in I$ and $(\lambda_k^\iota)_{k=1}^\infty$ is a subsequence of $\mathbf{N}$.
By our assumption, for each $\iota \in I$ there exists a linearly independent sequence $(x_k^\iota)_k$ in E_ι and a continuous seminorm q_ι on $S_\iota = sp((x_k^\iota)_k)$ such that

$$q_\iota\left(\sum_{k=1}^n \lambda_k x_k^\iota\right) = \sum_{k=1}^n \mid \lambda_k \mid$$

for all finite sequences $\lambda_1, ..., \lambda_n$. For $k \in \mathbf{N}$ let $y_k := (\lambda_k^\iota x_k^\iota)_{\iota \in I} \in F$ and $S := sp((y_k)_k)$. If q is an arbitrary seminorm on S we can choose $\iota_o \in I$ so that $\lambda_k^{\iota_o} \geq q(y_k)$ for each $k \in \mathbf{N}$. Let π_{ι_o} be the canonical projection of E onto E_{ι_o}. Then $\overline{q}_{\iota_o} := q_{\iota_o} \circ \pi_{\iota_o}$ is a continuous seminorm on S and

$$\overline{q}_{\iota_o}\left(\sum_{k=1}^n a_k y_k\right) = q_{\iota_o}\left(\sum_{k=1}^n a_k \lambda_k^{\iota_o} x_k^{\iota_o}\right) =$$

$$= \sum_{k=1}^{n} | a_k | \lambda_k^{\iota_0} \geq \sum_{k=1}^{n} | a_k | q(y_k) \geq q\left(\sum_{k=1}^{n} a_k y_k\right).$$

Hence q is dominated on S by $\overline{q}_{\iota_0}$, implying that q itself is continuous on S. Since q was arbitrarily chosen, S has the strongest locally convex topology.
Moreover, $(y_k)_k$ is linearly independent: If $\sum_{k=1}^{n} \mu_k y_k = 0$ then for each $\iota \in I$ we have $\sum_{k=1}^{n} \mu_k \lambda_k^{\iota} x_k^{\iota} = 0$, so that $\mu_k \lambda_k^{\iota} = 0$ for $k \leq n$ and each $\iota \in I$ ($(x_k^{\iota})_k$ is linearly independent). But then necessarily $\mu_k = 0$ for $k \leq n$.
Therefore S is isomorphic to Φ.
(7) $\Rightarrow$ (7′) and (6) $\Rightarrow$ (6′) are clear.
(7) $\Rightarrow$ (6) and (7′) $\Rightarrow$ (6′) follow from the fact that Φ itself is Φ-productive:
Choose a Hamel basis $\{x_n \mid n \in \mathbf{N}\}$ of Φ and set $p : \Phi \to \mathbf{R}^+$, $p(\sum_{i=1}^{n} \lambda_i x_i) := \sum_{i=1}^{n} | \lambda_i |$. Then p is a seminorm on Φ and therefore continuous.
(6′) $\Rightarrow$ (4) Assume that E is isomorphic to a subspace of $\mathbf{K}^J$. Then E^I is isomorphic to a subspace of $(\mathbf{K}^J)^I = \mathbf{K}^{J \times I}$. Applying (5) $\Rightarrow$ (4) to E^I now shows E^I not to be Φ-productive which contradicts (6′). ■

2.3.4 Example Let $(s) := \{(x_n)_n \mid \sup_n \mid n^k x_n \mid < \infty \text{ for } k = 1, 2, ...\}$ be the space of all rapidly decreasing sequences. Together with the generating system of seminorms $\{p_k((x_n)_n) = \sup_n \mid n^k x_n \| \, k \in \mathbf{N}\}$, (s) is a nuclear Fréchet space. Since each p_k is actually a norm, condition (3) of 2.3.3 is satisfied. Hence (s) is Φ-productive.
A deep result by T. and Y.KOMURA ([27], p.287) shows that an lcs is nuclear if and only if it is isomorphic to a subspace of $(s)^I$ for some indexing set I. By 2.3.3, $(s)^I$ contains Φ for each I with $| I | \geq 2^{\aleph_0}$. Thus many non-metrizable nuclear spaces contain Φ.

2.4 References

The results presented in 2.1 are based on standard material that can be found in any textbook on locally convex spaces (e.g. [23], [50], or [24]).
2.2.2 is a slightly improved version of a theorem by S.A.SAXON ([45]). Part (i) of 2.2.2 was already known to N.BOURBAKI ([6], section 2, chapter III, Ex. 4 (b)).
The definition of Φ-productive spaces and the main result 2.3.3 are also taken from [45].

3 Varieties of Locally Convex Spaces

The theory of varieties of locally convex spaces, initiated by J.DIESTEL, S.A.MORRIS and S.A.SAXON in [12] and [13] constitutes a basic setting in which several results contained in the present treatise can be stated in utmost generality. Apart from this it appears to us as a fruitful and interesting theory by itself. This chapter is intended to provide some of its fundamental theorems and to show the main directions of research it has given rise to.

3.1 Construction of Varieties

3.1.1 Definition *A nonempty class $\mathcal{V}$ of lcs's is called a* variety *if it is closed under the operations of taking subspaces, separated quotients, cartesian products and isomorphic images.*

3.1.2 Examples

(i) The class of all lcs's.

(ii) The class of all Schwartz spaces. ([23], p.278/279)

(iii) The class $\mathcal{N}$ of all nuclear spaces. ([50], III.7.4)

(iv) The class of all lcs's having their weak topologies. ([23], ch.3)

Since projective limits can be viewed as subspaces of products, each variety is closed under the operation of taking projective limits. On the other hand, a variety is not necessarily closed under the operations of taking inductive limits or locally convex direct sums.

3.1.3 Definition *For an arbitrary class $\mathcal{C}$ of lcs's let $\mathcal{V}(\mathcal{C})$ denote the intersection of all varieties containing $\mathcal{C}$. Then $\mathcal{V}(\mathcal{C})$ is said to be the* variety generated by $\mathcal{C}$. *If $\mathcal{C}$ consists of a single lcs E, then $\mathcal{V}(E) = \mathcal{V}(\mathcal{C})$ is called* singly generated.

In the sequel, we denote by (i) $S(\mathcal{C})$, (ii) $Q(\mathcal{C})$, (iii) $C(\mathcal{C})$ and (iv) $P(\mathcal{C})$ the class of all lcs's isomorphic to

(i) subspaces of elements of $\mathcal{C}$,

(ii) separated quotients of elements of $\mathcal{C}$,

(iii) cartesian products of families of lcs's in $\mathcal{C}$ and

(iv) products of finite families of lcs's in $\mathcal{C}$

The following theorems will show that a finite number of applications of the operations defined above always suffices to obtain $\mathcal{V}(\mathcal{C})$ from $\mathcal{C}$. In order to shorten the proofs of these results we first state some auxiliary results.

3.1.4 Lemma *Let E,F be tvs's and $f : E \to F$ a linear, continuous and open surjection. Then the associated injection $\hat{f} : E/f^{-1}(0) \to F$, $\hat{f}(x + f^{-1}(0)) = f(x)$ is a topological isomorphism.*

Proof. Clearly $\hat{f}$ is an algebraical isomorphism. Denote by $\pi : E \to E/f^{-1}(0)$ the canonical projection. Then $\hat{f}$ is continuous by the universal property of π because $\hat{f} \circ \pi = f$. Finally, $\hat{f}$ is open since the images of all neighborhoods of 0 in E under π form a neighborhood base of 0 in $E/f^{-1}(0)$ and since f is open. ■

3.1.5 Lemma *Let M be a subspace of the tvs E, $\pi : E \to E/M$ the canonical quotient map and let G' be a subspace of E/M. Then $G/M \cong G'$ topologically, where $G = \pi^{-1}(G')$.*

Proof. Set $\pi_o := \pi |_G$. Then $\pi_o : G \to G'$ is surjective, linear and continuous. Moreover, $\pi_o^{-1}(0) = \pi^{-1}(0) = M$ since $M \subseteq \pi^{-1}(G')$.
By 3.1.4 it remains to show that π_o is open: Let $H = H_o \cap G$ be open in G, where H_o is open in E. Then $\pi(H) = \pi(H_o \cap \pi^{-1}(G')) = \pi(H_o) \cap G'$, which is open in G' because $\pi(H_o)$ is open in E/M. ■

3.1.6 Lemma *Let E be a tvs and G,M subspaces of E with $M \subseteq G$. Then*

$$(E/M) / (G/M) \cong E/G$$

topologically.

Proof. We consider G/M as a topological subspace of E/M. Set $f : E/M \to E/G$, $f(x+M) = x+G$. f is linear and surjective and $f^{-1}(0) = G/M$. Let $\pi_1 : E \to E/M$ and $\pi_2 : E \to E/G$ be the canonical projections. Since $f \circ \pi_1 = \pi_2$ it follows that f is continuous and open (cf. the proof of 3.1.4).

The result now follows from 3.1.4. ∎

3.1.7 Theorem *For an arbitrary class $\mathcal{C}$ of lcs's, $\mathcal{V}(\mathcal{C}) = QSC(\mathcal{C})$.*

Proof. We only have to show that $QSC(\mathcal{C})$ is a variety. To do this, we succesively prove the following statements (which we also will repeatedly need in the course of this chapter):

(i) $SS(\mathcal{C}) = S(\mathcal{C})$

(ii) $CC(\mathcal{C}) = C(\mathcal{C})$

(iii) $QQ(\mathcal{C}) = Q(\mathcal{C})$

(i) and (ii) are trivial. For the proof of (iii) let $E \in \mathcal{C}$, M a closed subspace of E and G' a closed subspace of E/M. Then for some closed subspace $G \supseteq M$ of E, $G' = G/M$ by 3.1.5 and $(E/M)/(G/M) \cong E/G$ by 3.1.6. Hence $QQ(\mathcal{C}) \subseteq Q(\mathcal{C})$. Conversely, $E/M \cong (E/M)/(\{0\}/\{0\})$.

(iv) $SC(\mathcal{C}) \supseteq CS(\mathcal{C})$ is obvious.

(v) $QC(\mathcal{C}) \supseteq CQ(\mathcal{C})$:

Let I be any indexing set and for $i \in I$ let M_i be a closed subspace of $E_i \in \mathcal{C}$. Denote by $p_\iota : E_\iota \to E_\iota/M_\iota$ the canonical projections and let

$$f : \prod_{\iota\in I} E_\iota \to \prod_{\iota\in I} (E_\iota/M_\iota)\,,\; f(x) = (p_\iota(x_\iota))_{\iota\in I}\,.$$

f is continuous and open and $f^{-1}(0) = \prod_{\iota\in I} M_\iota$, so that

$$\prod_{\iota\in I} (E_\iota/M_\iota) \cong \left(\prod_{\iota\in I} E_\iota\right) \Big/ \left(\prod_{\iota\in I} M_\iota\right)$$

by 3.1.4.

(vi) $SQ(\mathcal{C}) = QS(\mathcal{C})$:

Let $E \in \mathcal{C}$, M a closed subspace of E and G' a subspace of E/M. Then for some subspace G of E, $G \supseteq M$ and $G' = G/M$ (cf. 3.1.5). Hence $SQ(\mathcal{C}) \subseteq QS(\mathcal{C})$. The reverse inclusion is self-evident. Thus we have:

(a). $Q(QSC(\mathcal{C})) = QSC(\mathcal{C})$, by (iii).

(b) $S(QSC(\mathcal{C})) = QSSC(\mathcal{C})$, by (vi); $= QSC(\mathcal{C})$, by (i).

(c) $C(QSC(\mathcal{C})) \subseteq QCSC(\mathcal{C})$, by (v); $\subseteq QSCC(\mathcal{C})$, by (iv); $= QSC(\mathcal{C})$, by (ii).

■

3.1.8 Corollary *If $\mathcal{C}$ is any class of lcs's, then $\mathcal{V}(\mathcal{C}) = SQC(\mathcal{C})$.*

Proof. $\mathcal{V}(\mathcal{C})) = QSC(\mathcal{C}) = SQC(\mathcal{C})$, by (vi). ■

In order to proof the following lemma, we introduce some notations:
For an arbitrary family $\{E_\iota \mid \iota \in I\}$ of lcs's let $\mathcal{F}$ be the system of all nonempty finite subsets of I ordered by inclusion. If $\sigma, \tau \in \mathcal{F}$, $\sigma \leq \tau$, then π_σ resp. $\pi_{\sigma\tau}$ denote the natural projections from $E := \prod_{\iota \in I} E_\iota$ onto $E_\sigma := \prod_{\iota \in \sigma} E_\iota$ resp. from E_τ onto E_σ.

3.1.9 Lemma *With E, $\{E_\iota \mid \iota \in I\}$ as above, let M be a closed subspace of E. For each $\sigma \in \mathcal{F}$ let M_σ be the closure of $\pi_\sigma(M)$ in E_σ.*
For $\sigma \leq \tau$ define $f_{\sigma\tau} : E_\tau/M_\tau \to E_\sigma/M_\sigma$ by $f_{\sigma\tau}(x + M_\tau) = \pi_{\sigma\tau}(x) + M_\sigma$.
Then $(E_\sigma/M_\sigma, f_{\sigma\tau})$ is a projective system whose projective limit F contains a dense subspace F_o isomorphic to E/M.

Proof. $\pi_{\sigma\tau}(M_\tau) \subseteq M_\sigma$ for $\tau \geq \sigma$ since $\pi_{\sigma\tau}$ is continuous. Hence $f_{\sigma\tau}$ is well-defined. For $\sigma \in \mathcal{F}$ let $p_\sigma : E_\sigma \to E_\sigma/M_\sigma$ be the canonical epimorphism. Then $f_{\sigma\tau} \circ p_\tau = p_\sigma \circ \pi_{\sigma\tau}$ for $\tau \geq \sigma$, implying that $f_{\sigma\tau}$ is continuous. Hence $(E_\sigma/M_\sigma, f_{\sigma\tau})$ is a projective system whose projective limit F is the subspace of $\prod_{\sigma \in \mathcal{F}} E_\sigma/M_\sigma$ formed by those vectors $(y_\sigma)_{\sigma \in \mathcal{F}}$ satisfying $f_{\sigma\tau}(y_\tau) = y_\sigma$ for $\sigma \leq \tau$.
F contains the subspace $F_o := \{(p_\sigma\pi_\sigma(x))_{\sigma \in \mathcal{F}} \mid x \in E\}$ and we claim that F_o is dense in F: Let $(y_\sigma)_{\sigma \in \mathcal{F}} \in F$ and $\sigma_1, ..., \sigma_n \in \mathcal{F}$. Define $\tau := \bigcup_{k=1}^n \sigma_k$ and choose $x \in E$ in such a way that $p_\tau\pi_\tau(x) = y_\tau$. Then for $k = 1, ..., n$, $p_{\sigma_k}\pi_{\sigma_k}(x) = f_{\sigma_k\tau}p_\tau\pi_\tau(x) = f_{\sigma_k\tau}(y_\tau) = y_{\sigma_k}$. Thus $u := (p_\sigma\pi_\sigma(x))_{\sigma \in \mathcal{F}} \in F_o$ and $y - u$ vanishes on the coordinates $\sigma_1, ..., \sigma_n$, so that F_o is dense in F.
It remains to show that E/M is isomorphic to F_o.
The family of mappings $p_\sigma\pi_\sigma : E \to E_\sigma/M_\sigma$ induces a continuous linear map $h_o : E \to \prod_{\sigma \in \mathcal{F}}(E_\sigma/M_\sigma)$. h_o, vanishing on M and having F_o as its image, gives rise to a continuous linear surjection $h : E/M \to F_o$. In order to show that h is open (and thus a quotient map by 3.1.4) we will prove that $h_o(U)$ is a 0-neighborhood in F_o for each open 0-neighborhood U in E.
By definition of the product topology it is sufficient to consider the case $U =$

$\pi_\tau^{-1}(U_\tau)$, where U_τ is an open 0-neighborhood in E_τ for some $\tau \in \mathcal{F}$. Define $V := \{(y_\sigma)_{\sigma\in\mathcal{F}} \in \prod_{\sigma\in\mathcal{F}}(E_\sigma/M_\sigma) \mid y_\tau \in p_\tau(U_\tau)\}$. Obviously, V contains 0 and is open since p_τ is open. We are going to show $V \cap F_o \subseteq h_o(U)$, thereby establishing $h_o(U)$ to be a 0-neighborhood in F_o.
First, observe that $U \cap (E \setminus (M+U)) = \emptyset$. Now it is crucial for the proof that $U = \pi_\tau^{-1}(U_\tau)$ and $(E \setminus (M+U))$ remain disjoint even after application of $p_\tau\pi_\tau$, i.e. we claim

$$p_\tau(U_\tau) \cap p_\tau\pi_\tau(E \setminus (M+U)) = \emptyset \qquad (*)$$

To see this, consider $x \in E \setminus (M+U)$. $x \notin M+U$ implies $(x+M) \cap U = \emptyset$. Since $U = \pi_\tau^{-1}(U_\tau)$, this gives $(\pi_\tau(x)+\pi_\tau(M)) \cap U_\tau = \emptyset$. Taking into account $M_\tau = \overline{\pi_\tau(M)}$ we even get $(\pi_\tau(x) + M_\tau) \cap U_\tau = \emptyset$ which, finally, results in $p_\tau\pi_\tau(x) \notin p_\tau(U_\tau)$, as desired.
Now let $h_o(x) = (p_\sigma\pi_\sigma(x))_\sigma \in F_o \cap V$. By definition of V, $p_\tau\pi_\tau(x) \in p_\tau(U_\tau)$. From $(*)$ it follows that $x \in M+U$, i.e. $x = m+u$ where $m \in M$ and $u \in U$. As $p_\sigma\pi_\sigma(m) = 0$ for all σ, $h_o(u) = h_o(x) \in h_o(U)$ which establishes $F_o \cap V \subseteq h_o(U)$.
To complete the proof that h is an isomorphism it remains to show that h is injective. Fortunately, this is already contained implicitly in the above argument: If $p : E \to E/M$ denotes the canonical quotient map, let $0 \neq \hat{x} \in E/M$, $\hat{x} = p(x)$, $x \in E$. Choose an open neighborhood U of 0 in E (which, as above, can be assumed to be of the form $U = \pi_\tau^{-1}(U_\tau)$) such that $p(U)$ does not contain $\hat{x} = p(x)$, i.e. $x \notin M+U$. From $(*)$ we obtain $p_\tau\pi_\tau(x) \notin p_\tau(U_\tau)$. This implies $h(\hat{x})_\tau = h_o(x)_\tau = p_\tau\pi_\tau(x) \neq 0$ and, hence, $h(\hat{x}) \neq 0$. ■

3.1.10 Theorem *For an arbitrary class $\mathcal{C}$ of lcs's, $\mathcal{V}(\mathcal{C}) = SCQP(\mathcal{C})$.*

Proof. 3.1.9 shows that $QC(\mathcal{C}) \subseteq SSCQP(\mathcal{C})$. Therefore, by 3.1.8 we have $\mathcal{V}(\mathcal{C}) = SQC(\mathcal{C}) \subseteq SSSCQP(\mathcal{C}) = SCQP(\mathcal{C}) \subseteq \mathcal{V}(\mathcal{C})$. ■

3.1.11 Theorem *If $\mathcal{C}$ is any class of Fréchet spaces and $E \in \mathcal{V}(\mathcal{C})$, then the completion $\widetilde{E}$ of E is in $\mathcal{V}(\mathcal{C})$.*

Proof. Finite products and quotients of Fréchet spaces are again Fréchet, so that by 3.1.10 E is a subspace of a product F of Fréchet spaces. Since F is complete it follows that $\widetilde{E} = \overline{E}^F \subseteq F$. Hence $\widetilde{E} \in S(F) \subseteq SCQP(\mathcal{C}) = \mathcal{V}(\mathcal{C})$. ■

3.2 $T(m)$-Spaces

3.2.1 Definition *Let E be an lcs and m any infinite cardinal. If every 0-neighborhood in E contains a subspace of E of codimension strictly less than m, then E is called a $T(m)$-space.*

3.2.2 Remarks

(i) Since in every lcs each 0-neighborhood contains a closed 0-neighborhood, we can additionally require the subspace in 3.2.1 to be closed.

Now let E be an lcs of Hamel dimension n. Then

(ii) For all $m > n$, E is a $T(m)$-space ($codim(\{0\}) = dim(E)$).

(iii) If E is normable, then E is a $T(m)$-space if and only if $m > n$ (since $\{0\}$ is the only subspace contained in $\{x \mid \|x\| < 1\}$).

(iv) E is a $T(\aleph_0)$-space if and only if E has its weak topology (cf.2.3.3).

(v) It is easily verified that subspaces, products and continuous linear images (in particular: quotients) of $T(m)$-spaces are $T(m)$-spaces.

3.2.3 Theorem *If $\mathcal{C}$ is a class of $T(m)$-spaces, then $\mathcal{V}(\mathcal{C})$ contains only $T(m)$-spaces.*

Proof. 3.1.7 and 3.2.2 (v). ∎

3.2.4 Corollary *If E is an infinite-dimensional normed space and F is an lcs of smaller Hamel dimension, then $E \notin \mathcal{V}(F)$. In particular, $\mathcal{V}(E) \neq \mathcal{V}(F)$.*

Proof. Let $m := dim(E)$. By 3.2.2 (ii) and (iii), F is a $T(m)$-space but E is not. It follows that $E \notin \mathcal{V}(F)$. ∎

3.2.5 Corollary *The class of all varieties is not a set.*

Proof. For an arbitrary infinite cardinal m let $E := \mathbf{K}^{(m)}$. If $x = (x_\iota)_{\iota \in m} \in E$, define $\|x\| := \sum_{\iota \in m} \mid x_\iota \mid$. Then $(E, \| \ \|)$ is a normed space of Hamel dimension m. Hence there exist normed spaces of arbitrarily large dimension. By 3.2.4, then, there are at least as many different varieties as there are cardinal numbers. ∎

3.2.6 Corollary *The variety of all lcs's is not singly generated.* ∎

3.2.7 Definition *For each infinite cardinal m let $\mathcal{V}_m$ denote the variety of all $T(m)$-spaces.*

We note that $\mathcal{V}_m \subseteq \mathcal{V}_n$ for $m \leq n$.

3.2.8 Proposition *Let m be an infinite cardinal and $E \in \mathcal{V}_m$.*
Then $E \in SC(\mathcal{C})$, where $\mathcal{C}$ denotes the class of all normed linear spaces having Hamel dimension strictly less than m (or, equivalently, being a member of $\mathcal{V}_m$ - cf. 3.2.2 (iii)).

Proof. Let $\{U_\iota \mid \iota \in I\}$ be a base of absolutely convex neighborhoods of 0 in E. Set $E_\iota := p_{U_\iota}^{-1}(0)$ (p_{U_ι} the gauge of U_ι) and equip $F_\iota := E/E_\iota$ with the norm $\|\hat{x}\| := p_{U_\iota}(x)$ ($\hat{x} = p_\iota(x)$, $x \in E$, $p_\iota : E \to E/E_\iota$ the canonical mapping, cf. [50], p.97).
Now $E \in \mathcal{V}_m \Rightarrow E$ is a $T(m)$-space $\Rightarrow F_\iota$ is a $T(m)$-space (3.2.2 (v)) $\Rightarrow dim(F_\iota) < m$ (3.2.2 (iii)).
We complete the proof by demonstrating that E is isomorphic to a subspace of $\prod_{\iota \in I} F_\iota$: $f : E \to \prod_{\iota \in I} F_\iota$, $f = (p_\iota)_{\iota \in I}$ is continuous. Let $x \neq 0$. Then for some $\iota \in I$, $x \notin U_\iota \supseteq E_\iota$. Hence $p_\iota(x) \neq 0$ and f is injective.
We claim that f is an isomorphism onto its image. Indeed, a straightforward computation shows that for every $\iota_o \in I$,

$$\left(p_{\iota_o}(U_{\iota_o}) \times \prod_{\iota \neq \iota_o} F_\iota \right) \cap f(E) \subseteq 2f(U_{\iota_o}),$$

so that $f(U_{\iota_o})$ is a neighborhood of 0 in $f(E)$. ■

3.2.9 Remark If U is an absolutely convex neighborhood of 0 in E, then $p_U^{-1}(0) = \{x \in E \mid \mathbf{K}x \subseteq U\}$ obviously is the largest subspace contained in U.

3.2.10 Theorem *A variety $\mathcal{V}$ is singly generated if and only if $\mathcal{V} \subseteq \mathcal{V}_m$ for some infinite m.*

Proof. Suppose $\mathcal{V} = \mathcal{V}(E)$ for some lcs E of Hamel dimension n. Then E is a $T(m)$-space for each $m > n$ (by 3.2.2 (ii)). Hence $\mathcal{V} \subseteq \mathcal{V}_m$ by 3.2.3.
Conversely, assume that $\mathcal{V} \subseteq \mathcal{V}_m$ for some m. Let $\mathcal{P}$ be the class of all lcs's in $\mathcal{V}$ of Hamel dimension strictly less than m. Fix a vector space F of Hamel dimension m. Let $\mathcal{B}$ be the set(!) of all linear subspaces of F equipped with all possible locally convex Hausdorff topologies and define $E := \prod\{G \mid G \in \mathcal{P} \cap \mathcal{B}\}$. (F only serves to

avoid the set-theoretic difficulties connected with $\widehat{E} := \prod\{G \mid G \in \mathcal{P}\}$; in a sense, E is just as good as $\hat{E}$ since each $G \in \mathcal{P}$ has an isomorphic copy in $\mathcal{P} \cap \mathcal{B}$.)
Now $S(E)$ contains an isomorphic copy of every $G \in \mathcal{P} \cap \mathcal{B}$, hence of every $G \in \mathcal{P}$. Thus we obtain $\mathcal{P} \subseteq S(E)$. For any $H \in \mathcal{V}$, on the other hand, $\mathcal{V}(H) \subseteq \mathcal{V}$. 3.2.8 implies $H \in SC(\mathcal{P})$ which, in turn, shows $\mathcal{V} \subseteq SC(\mathcal{P}) \subseteq \mathcal{V}(\mathcal{P}) \subseteq \mathcal{V}$.
Putting all this together, we obtain $\mathcal{V} = SC(\mathcal{P}) \subseteq SCS(E) \subseteq SSC(E) = SC(E) \subseteq \mathcal{V}(E) \subseteq \mathcal{V}$, so that $\mathcal{V} = SC(E) = \mathcal{V}(E)$ is singly generated. ■

3.2.11 Corollary *Each subvariety of a singly generated variety is singly generated.* ■

3.2.12 Definition *Let $\mathcal{V}$ be a variety and $E \in \mathcal{V}$. If each member of $\mathcal{V}$ is isomorphic to a subspace of a product of copies of E, then E is called a* universal generator *for $\mathcal{V}$.*

3.2.13 Theorem *Every singly generated variety has a universal generator.*

Proof. This follows from the last line of the proof of 3.2.10. ■

3.3 Φ-Productive Spaces and the Relative Size of Varieties

3.3.1 Proposition *Any variety generated by an infinite-dimensional normed linear space contains a maximal proper subvariety.*

Proof. Let N be an infinite-dimensional normed linear space. By 3.2.11 each subvariety of $\mathcal{V}(N)$ is singly generated. Set $\mathcal{C} := \{F \in \mathcal{V}(N) \mid \mathcal{V}(F) \subset \mathcal{V}(N)\}$. Then $\mathcal{C} \neq \emptyset$ since $\{0\} \in \mathcal{C}$ and a subvariety of $\mathcal{V}(N)$ is proper if and only if it is generated by an element of $\mathcal{C}$.
Define a partial order on $\mathcal{C}$ by $E_1 \leq E_2$ if $\mathcal{V}(E_1) \subseteq \mathcal{V}(E_2)$ $(E_1, E_2 \in \mathcal{C})$. Suppose that Υ is an ascending chain in $\mathcal{C}$. If $N \in \mathcal{V}(\Upsilon)$ then $N \in SQP(\Upsilon)$ by 3.4.5 below. Thus for some $E_1, ..., E_n \in \Upsilon$, $E_i \leq E_n$ $(1 \leq i \leq n)$ we have $N \in SQ(E_1 \times ... \times E_n)$. This implies $N \in \mathcal{V}(E_1 \times ... \times E_n) = \mathcal{V}(E_n)$, contradicting the fact that $\mathcal{V}(E_n)$ is a proper subvariety of $\mathcal{V}(N)$.
Hence $N \notin \mathcal{V}(\Upsilon)$ and $\mathcal{V}(S) = \mathcal{V}(F)$ for some $F \in \mathcal{C}$. That is to say, F is an upper bound for Υ. By Zorn's lemma there exists a maximal element M in $\mathcal{C}$, i.e. $\mathcal{V}(M)$ is a maximal proper subvariety of $\mathcal{V}(N)$. ■

The following results reveal the importance of the spaces Φ_m introduced in chapter

2 for questions concerning the relative size of varieties.

3.3.2 Theorem *Let n be any infinite cardinal and m the smallest cardinal greater than n. Then*

(i) Φ_n is a universal generator for $\mathcal{V}(\Phi_n)$.

(ii) $\mathcal{V}(\Phi_n)$ is the unique maximal proper subvariety of $\mathcal{V}(\Phi_m)$.

(iii) $\mathcal{V}(\Phi_m) \cap \mathcal{V}_m = \mathcal{V}(\Phi_n)$.

Proof. By 3.1.10, $\mathcal{V}(\Phi_n) = SCQP(\Phi_n)$. In 2.1 we showed that each member of $P(\Phi_n)$ is isomorphic to Φ_n and that each quotient of Φ_n is isomorphic to $\Phi_k \in S(\Phi_n)$ for some $k \leq n$. It follows that $\mathcal{V}(\Phi_n) = SCQ(\Phi_n) \subseteq SCS(\Phi_n) \subseteq SSC(\Phi_n) = SC(\Phi_n)$, implying (i).
Let E be a member of a subvariety $\mathcal{V}$ of $\mathcal{V}(\Phi_m)$. By (i), E is isomorphic to a subspace of a product $\prod_{\iota\in I} F_\iota$, where $F_\iota = \Phi_m$ for each $\iota \in I$. Then $E_\iota := \pi_\iota(E)$ is isomorphic to Φ_{k_ι} for some $k_\iota \leq m$ ($\iota \in I$). If each $k_\iota < m$, then $k_\iota \leq n$ and $E \in SC(\Phi_n) = \mathcal{V}(\Phi_n)$.
On the other hand, if $k_\iota = m$ for some ι, then π_ι is a continuous linear mapping from E onto Φ_m. By continuity, the topology of Φ_m is coarser than the quotient topology. Consequently, the topology of Φ_m - being the strongest of all locally convex topologies on Φ_m - coincides with the quotient topology, i.e. E has a quotient isomorphic to Φ_m and therefore $\mathcal{V}(E) \subseteq \mathcal{V} \subseteq \mathcal{V}(\Phi_m) \subseteq \mathcal{V}(E)$, so that $\mathcal{V} = \mathcal{V}(\Phi_m)$.
Thus either $\mathcal{V} \subseteq \mathcal{V}(\Phi_n)$ (if $k_\iota < m$ for all $E \in \mathcal{V}$) or $\mathcal{V} = \mathcal{V}(\Phi_m)$ (otherwise).
Next we observe that $\Phi_m \notin \mathcal{V}_m$: Let B be any Hamel basis of Φ_m. Then $\{0\}$ is the only subspace contained in the 0-neighborhood

$$U := \left\{ x \in \Phi_m \mid x = \sum_{b\in B} x_b b \ , \ | x_b | \leq 1 \ \forall b \in B \right\}.$$

But clearly $\Phi_n \in \mathcal{V}_m$. By 3.2.3, $\mathcal{V}(\Phi_n) \subset \mathcal{V}(\Phi_m)$ and (ii) follows.
Finally, $\Phi_n \in \mathcal{V}(\Phi_m) \cap \mathcal{V}_m$ so that $\mathcal{V}(\Phi_n) \subseteq \mathcal{V}(\Phi_m) \cap \mathcal{V}_m$. Conversely, since $\Phi_m \notin \mathcal{V}_m$, $\mathcal{V}(\Phi_m) \cap \mathcal{V}_m \subset \mathcal{V}(\Phi_m)$. Thus, by (ii), $\mathcal{V}(\Phi_m) \cap \mathcal{V}_m \subseteq \mathcal{V}(\Phi_n)$. ■

3.3.3 Theorem

(i) $\mathbf{K}$ is a universal generator for $\mathcal{V}^w$, the variety of all lcs's having their weak topology. In particular, $\mathcal{V}^w$ is the smallest nontrivial variety.

(ii) Each variety properly containing $\mathcal{V}(\mathbf{K})$ also contains $\mathcal{V}(\Phi)$, i.e. $\mathcal{V}(\Phi)$ is the unique second smallest variety.

(iii) There is no third smallest variety.

Proof. By 3.1.10, $\mathcal{V}(\mathbf{K}) = SCQP(\mathbf{K}) = SC(\mathbf{K})$. Moreover, an lcs has its weak topology if and only if it is a member of of $SC(\mathbf{K})$ (see 2.3.3). Thus $\mathcal{V}(\mathbf{K}) = \mathcal{V}^w$. It follows that $\mathbf{K}$ is a universal generator for $\mathcal{V}^w$. If $\mathcal{V}$ is any nontrivial variety, then it contains a 1-dimensional lcs E which is necessarily isomorphic to $\mathbf{K}$, so that $\mathcal{V}(\mathbf{K}) = \mathcal{V}(E) \subseteq \mathcal{V}$.

If $\mathcal{V}$ is a variety properly containing $\mathcal{V}(\mathbf{K})$ then there exists a member E of $\mathcal{V}$ which does not have its weak topology. Again by 2.3.3 it follows that E^I contains Φ for each indexing set I with $|I| \geq 2^{\aleph_0}$. Hence $\mathcal{V}(\Phi) \subseteq \mathcal{V}(E) \subseteq \mathcal{V}$.

For the proof of (iii) it is enough to construct two varieties strictly larger than $\mathcal{V}(\Phi)$ whose intersection is $\mathcal{V}(\Phi)$. Let $\aleph_1$ be the smallest cardinal greater than $\aleph_0$. By 3.2.2 (ii) and 3.2.3, $\mathcal{V}(\Phi) \subseteq \mathcal{V}_{\aleph_1}$ but $\Phi_{\aleph_1} \notin \mathcal{V}_{\aleph_1}$ (cf. the proof of 3.2.3 (iii)). Thus $\mathcal{V}(\Phi_{\aleph_1}) \supset \mathcal{V}(\Phi)$.

From the proof of 3.2.3 (ii) we infer that if $E \in \mathcal{V}(\Phi)$ then either $E \in \mathcal{V}^w$ or some quotient of E is isomorphic to Φ.

Now let E be a normed, $\aleph_0$-dimensional space. Then E does neither have its weak topology (cf. 2.3.3 (2)) nor is any quotient of E isomorphic to Φ since Φ is not metrizable. By the preceding argument, $E \in \mathcal{V}_{\aleph_1} \setminus \mathcal{V}(\Phi)$ and therefore $\mathcal{V}(\Phi) \subset \mathcal{V}_{\aleph_1}$. Finally, by 3.3.2 (iii), $\mathcal{V}_{\aleph_1} \cap \mathcal{V}(\Phi_{\aleph_1}) = \mathcal{V}(\Phi)$. ■

In particular, 3.3.3 implies that the mutually dual spaces ω and Φ are universal generators for the first and second smallest varieties, respectively.

3.3.4 Corollary *The statements of 2.3.3 are equivalent to*

(8) $\Phi \in \mathcal{V}(E)$.

Proof.
(7) $\Rightarrow$ (8) Clear.
(8) $\Rightarrow$ (1) Suppose $E \in \mathcal{V}^w$. Then $\mathcal{V}(\Phi) \subseteq \mathcal{V}(E) \subseteq \mathcal{V}^w$. But Φ does not have its weak topology (by (7) $\Rightarrow$ (1)). ■

The corollary also follows from 3.3.3 since E has its weak topology if and only if $\mathcal{V}(E)$ is contained in the smallest variety $\mathcal{V}^w = \mathcal{V}(\mathbf{K})$.

3.4 Applications

In the last part of this chapter we are going to apply the results obtained so far to a number of spaces important in functional analysis, e.g. the classical Banach spaces. To this end, we need several facts from the literature (see [13] for a very comprehensive bibliography):

(i) l^p is isomorphic to a subspace of $L^p[0,1]$ for $1 < p < \infty$.

(ii) Every separable Banach space is isomorphic to

  (a) a subspace of $C[0,1]$,

  (b) a quotient of l^1, and

  (c) a quotient of L^1.

(iii) l^1, $C[0,1]$ and $L^p[0,1]$ for $1 \leq p < \infty$, are separable Banach spaces.

(iv) If K is any uncountable compact metric space, then $C(K)$ is isomorphic to $C[0,1]$.

(v) l^1 is isomorphic to a subspace of l^∞.

(vi) l^∞ is isomorphic to L^∞.

From this we immediately conclude

3.4.1 Theorem *If K is any uncountable compact metric space and $1 < p < \infty$, then*

$$\mathcal{V}(l^p) \subseteq \mathcal{V}(L^p) \subseteq \mathcal{V}(l^1) = \mathcal{V}(C[0,1]) = \mathcal{V}(C(K)) = \mathcal{V}(L^1) \subseteq \mathcal{V}(l^\infty) = \mathcal{V}(L^\infty)$$

■

A straightforward modification of [50], II.5.4, Cor.2 shows that every separable lcs is isomorphic to a subspace of a product of separable Banach spaces. Let $\mathcal{C}$ denote the class of all separable Banach spaces. By (ii) (b) $\mathcal{C} = Q(l^1)$ so that $\mathcal{V}(l^1)$ contains all separable lcs's. Furthermore $\mathcal{C} \subseteq S(C[0,1])$ by (ii) (a) and $C[0,1] \in \mathcal{C}$ by (iii). Thus $SC(C[0,1]) = SSC(C[0,1]) \supseteq SCS(C[0,1]) \supseteq SC(\mathcal{C}) = SCQ(l^1) = SCQP(l^1)$ $= \mathcal{V}(l^1)$. (Concerning the fifth step in this chain, observe that $Q(l^1)$ contains all separable Banach spaces, hence, in particular, $QP(l^1)$. It follows that $Q(l^1) \subseteq$

$QP(l^1) \subseteq Q(l^1)$.) This, together with 3.4.1 proves

3.4.2 Corollary *$C[0,1]$ is a universal generator for $\mathcal{V}(l^1)$.* ■

Every Schwartz space is isomorphic to a subspace of a product of separable Banach spaces (see [52], 2., statement 9). Therefore

3.4.3 Corollary *$\mathcal{V}(l^1)$ contains the variety of all Schwartz spaces.* ■

An lcs E is nuclear (resp. a Schwartz space) if and only if every continuous linear map from E into any Banach space F is nuclear (resp. compact) (see [24], 17.1.7 and [50], III.7.2). Since every nuclear map is compact by [50], III.7.1, Cor.1, every nuclear lcs is a Schwartz space. From 3.2.11 and 3.2.13 we conclude

3.4.4 Theorem *The variety of all Schwartz spaces and therefore also the variety of all nuclear spaces both have universal generators.* ■

By 2.3.4, (s) (the space of all rapidly decreasing sequences) is a universal generator for the nuclear variety.
If E is an lcs, then E, equipped with the topology of uniform convergence on the compact subsets of $(E', \beta(E', E))$ is called the Schwartz space associated with E and denoted by E_o. By [24], p.204, theorem 1, $(c_o)_o$ is a universal generator for the variety of all Schwartz spaces (c_o denoting, as usual, the Banach space of all null sequences in $\mathbf{K}$). By the way, the completion of $(c_o)_o$ is $(l^\infty, \mu(l^\infty, l^1))$ (cf.[24], p.206 and 1.9.7). Further examples of universal Schwartz spaces can be found in [40].
For an arbitrary class $\mathcal{C}$ of lcs's let $P_C(\mathcal{C})$ denote the class of all lcs's isomorphic to a countable product of members of $\mathcal{C}$.

3.4.5 Theorem *Let $\mathcal{C}$ be a class of lcs's and $E \in \mathcal{V}(\mathcal{C})$.*

(i) If E is normable, then $E \in SQP(\mathcal{C}) = QSP(\mathcal{C})$.

(ii) If E is metrizable, then $E \in SP_CQP(\mathcal{C})$.

Proof. (i) $\mathcal{V}(\mathcal{C}) = SCQP(\mathcal{C})$ by 3.1.10. Furthermore, $SQP(\mathcal{C}) = SQPP(\mathcal{C}) = SPQP(\mathcal{C})$. Hence it suffices to show that if the normed space E is a subspace of $\prod_{\iota \in I} E_\iota$ (E_ι an lcs for $\iota \in I$), then E is isomorphic to a subspace of the product space $\prod_{\iota \in \sigma} E_\iota$ for some finite subset σ of I.
The unit ball U of E contains a set of the form $V \cap E$, where $V = \prod_{\iota \in I} V_\iota$ with each V_ι a neighborhood of 0 in E_ι and $V_\iota = E_\iota$ for all ι not in some finite subset σ of I.

Let $\pi_\sigma : E \to E_\sigma := \prod_{\iota\in\sigma} E_\iota$ denote the natural projection.
π_σ is an isomorphism onto its image: Obviously, π_σ is linear and continuous. If $x \notin U$ then $x \notin V$ and $\pi_\sigma(x) \neq 0$. Hence π_σ is injective. Now let $y' \in (\prod_{\iota\in\sigma} V_\iota) \cap \pi_\sigma(E)$. There exists some $y \in E$ with $\pi_\sigma(y) = y'$, implying that $y \in V$. Thus $(\prod_{\iota\in\sigma} V_\iota) \cap \pi_\sigma(E) \subseteq \pi_\sigma(V \cap E) \subseteq \pi_\sigma(U)$ and π_σ is open.
(ii)Let $\{U_n \mid n \in \mathbf{N}\}$ be a neighborhood base of 0 in E. For each $n \in \mathbf{N}$ there exists $V_n = \prod_{\iota\in\sigma_n} V_\iota \times \prod_{\iota\notin\sigma_n} E_\iota$ (V_ι a neighborhood of 0 in E_ι for $\iota \in I$) such that $V_n \cap E \subseteq U_n$.
Define $\sigma := \bigcup_{n=1}^\infty \sigma_n$ and let $\pi_\sigma : E \to \prod_{\iota\in\sigma} E_\iota$ be the natural projection. Again, π_σ is linear and continuous. If $x \neq 0$ then for some $n \in \mathbf{N}$, $x \notin U_n$. Hence $x \notin V_n$, $\pi_{\sigma_n}(x) \neq 0$, $\pi_\sigma(x) \neq 0$ and π_σ is injective.
Finally, for each $n \in \mathbf{N}$ we have

$$\left(\prod_{\iota\in\sigma_n} V_\iota \times \prod_{\iota\in\sigma\setminus\sigma_n} E_\iota\right) \cap \pi_\sigma(E) \subseteq \pi_\sigma(V_n \cap E) \subseteq \pi_\sigma(U_n)$$

so that π_σ is open onto its image. ■

3.4.6 Corollary *Let $\mathcal{H}$ be a class of Hilbert spaces and B any Banach space in $\mathcal{V}(\mathcal{H})$. Then B is isomorphic as lcs to a Hilbert space.*

Proof. $QP(\mathcal{H})$ contains only spaces isomorphic to Hilbert spaces. By 3.4.5, $B \in SQP(\mathcal{H})$. Thus B is isomorphic to a complete and therefore closed subspace of a Hilbert space, i.e. to a Hilbert space. ■

3.4.7 Corollary *For $1 \leq p (\neq 2) \leq \infty$, $l^p \notin \mathcal{V}(l^2) = \mathcal{V}(L^2)$. Consequently, $L^p \notin \mathcal{V}(l^2)$.*

Proof. This follows from 3.4.6 and the fact that there is no lcs-isomorphism from l^p onto l^2 (see [34], p.31, I.2.7). ■

3.4.8 Proposition *Let $\mathcal{B}$ be a class of reflexive Banach spaces and $E \in \mathcal{V}(\mathcal{B})$. Then $\widetilde{E}$ is semireflexive. If, moreover, E is evaluable, then $\widetilde{E}$ is reflexive.*
In particular, any Fréchet space or strict (LF)-space in $\mathcal{V}(\mathcal{B})$ is reflexive.

Proof. $\widetilde{E} \in \mathcal{V}(\mathcal{B})$ by 3.1.11. By [50], IV.5.8 $CQP(\mathcal{B})$ contains only semireflexive lcs's. Furthermore, any complete subspace of a semireflexive lcs is semireflexive (cf. [50], p.144). The completion of an evaluable space is evaluable (cf. [24], 11.3.1). Since every evaluable, semireflexive lcs is reflexive, the proof is complete. ■

3.4.9 Corollary *For $1 < p < \infty$, neither c_o nor L^1 is in $\mathcal{V}(L^p)$. Hence $\mathcal{V}(L^p) \subset \mathcal{V}(L^1) = \mathcal{V}(l^1) = \mathcal{V}(C[0,1])$.*

Proof. 3.4.1 and 3.4.8. ■

A lot more can be said about which of the inclusions in 3.4.1 are proper. However, the proofs of such results make extensive use of the vast literature on classical Banach spaces. Therefore we only state a few of them and refer the interested reader to [13] for the details.

- *For $1 < p \neq q < \infty$, $l^p \notin \mathcal{V}(l^q)$.*

- *For $1 < p \neq q < \infty$, the following statements are equivalent:*

  (i) $L^p \in \mathcal{V}(L^q)$.

  (ii) $l^p \in \mathcal{V}(L^q)$.

  (iii) Either $q < p \leq 2$ or $2 \leq p < q$.

- *If $1 < p(\neq 2) < \infty$, then $\mathcal{V}(l^p) \subset \mathcal{V}(L^p)$.*

- *For any $1 \leq p \leq \infty$, $l^p \notin \mathcal{V}(c_o)$.*

- $\mathcal{V}(l^1) \subset \mathcal{V}(l^\infty)$.

- *If $1 < p < \infty$ then none of the spaces l^1, L^1, $C[0,1]$ is in any of the varieties $\mathcal{V}(c_o)$, $\mathcal{V}(l^p)$ or $\mathcal{V}(L^p)$.*

- *If B is any infinite-dimensional Banach space and $\mathcal{N}$ is the variety of all nuclear spaces, then $\mathcal{N} \subsetneq SC(B)$ (cf. [44]).*

3.5 References

Except for the remarks on universal Schwartz spaces following 3.3.4, all the results contained in chapter 3 are due to J.DIESTEL, S.A.MORRIS and S.A.SAXON ([12] and [13]).

4 Linear Baire Spaces

4.1 Characterizations

A subset A of a topological space X is called *rare* (or *nowhere dense*) in X if the closure of A has void interior ($int(\overline{A}) = \emptyset$). A is *meager* (or *of first category*) in X if A is the union of a sequence of rare subsets of X. Finally, a topological space X is a *Baire space* if no nonempty open subset of X is meager in X. Thus, if a Baire space X is the union of a sequence $(A_n)_{n\in\mathbf{N}}$ of closed subsets then at least one A_n has an interior point.

The properties of A being rare in X resp. meager in X obviously depend on the surrounding space X ('closure' and 'interior' both refer to X). However, if the underlying space is clear from the context, we will simply use the abbreviations 'rare' resp. 'meager'.

X is a Baire space if and only if the intersection of any countable family of dense open subsets of X is dense in X (cf. [51], p.133). It is well known that every locally compact Hausdorff space and every complete metrizable topological space is a Baire space (Baire's theorem, [51], p.134).

If, in addition to its topology, the space X is endowed with a compatible linear structure, i.e. if X is a tvs, then there are more useful criteria for X having the Baire property.

4.1.1 Proposition *A tvs E is a Baire space if and only if E cannot be covered by an arbitrary (resp. increasing) sequence of rare sets.*

Proof. The condition is clearly necessary.

Conversely, suppose E is not Baire. Then there exists an open subset $U \neq \emptyset$ of E which is meager. Since translation is a homeomorphism we may assume without loss of generality that $0 \in U$.

Because U is meager, there exists a sequence $(A_n)_n$ with $U = \bigcup_{n=1}^{\infty} A_n$ and $int(\overline{A}) = \emptyset$ for each n. But then

$$E = \bigcup_{m=1}^{\infty} mU = \bigcup_{m,n=1}^{\infty} mA_n,$$

so that E itself is covered by a sequence of rare sets.

The observation that finite unions of rare sets are rare completes the proof. ■

4.1.2 Lemma *Let (E, τ) be a one-dimensional tvs. If E is Hausdorff, then $E \cong \mathbf{K}$; otherwise $\tau = \{\emptyset, E\}$.*

Proof. $\overline{\{0\}}$ is a subspace of E. Hence $dim(\overline{\{0\}}) = 0$ or 1.

If $dim(\overline{\{0\}}) = 0$, then $\overline{\{0\}} = \{0\}$. Thus E is Hausdorff and the result follows from [50], I.3.1.

Otherwise $E = \overline{\{0\}}$. Let $\emptyset \neq A$ be a closed subset of E, $x \in A$. Then $0 \in -x + A$, so that $E = \overline{\{0\}} \subseteq -x + A$. Consequently, $A = E$. ■

In the sequel we will frequently use the fact that a topological space $X \neq \emptyset$ carrying the coarsest topology $\{\emptyset, X\}$ is a Baire space (the only dense open subset being X itself). A tvs carries the coarsest topology if and only if E is the only neighborhood of 0.

4.1.3 Theorem *A tvs (E, τ) is a Baire space if and only if every set B satisfying $E = \bigcup_{n=1}^{\infty}(nB)$ is not rare.*

Proof. Again, the necessity of the condition is trivial.

If E is not Baire, there exists a sequence $(A_n)_n$ of rare sets covering E. Since $\tau \neq \{\emptyset, E\}$, there is a balanced neighborhood U of 0 distinct from E.

Choose $x \in E \setminus U$ and V to be a closed, balanced neighborhood of zero such that $V + V \subseteq U$. Set $B_n := V \cap A_n$ and $B = \bigcup_{n=1}^{\infty} n^{-1}B_n$. Then we have

$$kV = k\bigcup_{n=1}^{\infty} B_n \subseteq k\bigcup_{n=1}^{\infty}(nB) \subseteq \bigcup_{n=1}^{\infty}(nB).$$

Thus $E = \bigcup_{n=1}^{\infty}(nB)$ since V is absorbing.

Suppose $\overline{B}$ contains some open neighborhood N of a point y. Then for each $k \in \mathbf{N}$

$$N \subseteq \overline{\bigcup_{n\in\mathbf{N}} n^{-1}B_n} = \overline{\bigcup_{n<k} n^{-1}B_n} \cup \overline{\bigcup_{n\geq k} n^{-1}B_n}.$$

Therefore $N \subseteq \overline{\bigcup_{n\geq k} n^{-1}B_n} \subseteq k^{-1}V$ because otherwise the rare set $\overline{\bigcup_{n<k} n^{-1}B_n}$ would contain a nonempty open set (namely $N \setminus \overline{\bigcup_{n\geq k} n^{-1}B_n}$).

Also, there is some $\delta > 0$ such that $y \pm \delta x \in N$. This implies that

$$2\delta x = (y + \delta x) - (y - \delta x) \in k^{-1}V + k^{-1}V \subseteq k^{-1}U$$

for each $k \in \mathbf{N}$. But since $2\delta x \notin k^{-1}U$ whenever $k^{-1} \leq 2\delta$, we arrive at a contradiction. Hence B is rare. ■

A straightforward argument shows that an lcs is barrelled if and only if there does not exist a rare, absolutely convex set B such that $E = \bigcup_{n=1}^{\infty}(nB)$ (cf. 1.4.14). 4.1.3 now yields that for locally convex spaces the term 'Baire' differs from 'barrelled' only in the omission of the property of being absolutely convex for the set B. Consequently, each locally convex Baire space is barrelled.
The reverse implication is, however, not valid. Actually, there are large classes of spaces which are barrelled but not Baire. A detailed analysis of these spaces will be given in chapters 5 and 6.
The following theorem (in its 'sufficiency part') is a strengthening of 4.1.3: If a tvs (E,τ) is not Baire, then the rare set B satisfying $E = \bigcup_{n=1}^{\infty}(nB)$ can even be chosen to be closed, balanced and absorbing.
Observe that $E = \bigcup_{n=1}^{\infty}(nB)$ is strictly weaker than B being absorbing as long as B is not balanced: Consider the set $B = B_+ \cup (-B_+)$, where

$$B_+ = \{0\} \cup \bigcup_{n=1}^{\infty} \left[\frac{1}{2^{2n}}, \frac{1}{2^{2n-1}}\right] \subseteq \mathbf{R};$$

Then $B \cup 2B = [-1,1]$, so that $\mathbf{R} = \bigcup_{n=1}^{\infty}(nB)$.

4.1.4 Definition *Let A be a subset of a tvs E. The* balanced core *of A is the largest balanced set contained in A and is denoted by $bc(A)$.*

For arbitrary A, $bc(A)$ really exists since any union of balanced sets is balanced. Moreover,

$$bc(A) = \{x \in A \mid \lambda x \in A \;\forall \mid \lambda \mid \leq 1\} = \bigcap_{|\lambda| \geq 1} \lambda A$$

whenever $bc(A) \neq \emptyset$ (cf. [23], p.80)

4.1.5 Theorem *A tvs (E,τ) is a Baire space if and only if every absorbing, balanced, closed set B satisfying $E = \bigcup_{n=1}^{\infty}(nB)$ is not rare.*

Proof. If B is absorbing, balanced and rare, then $E = \bigcup_{n=1}^{\infty}(nB)$ is meager so that E is not Baire.
Now suppose E is not Baire. As in the proof of 4.1.3 we can find a balanced neighborhood U of 0 and $x \in E \setminus U$. Let V be a closed, balanced 0-neighborhood with $V + V \subseteq U$.

According to 4.1.1, there exists an increasing sequence of rare sets $(C_n)_n$ covering E. The sets $A_n := \overline{C_n} \cap V$ are rare, closed and their union is V.
If E is a complex vector space, define

$$B_n := \bigcup_{k=0}^{n-1} e^{\frac{2k\pi i}{n}} A_n$$

If E is real, set

$$B_n := A_n \cup (-A_n).$$

The rest of the proof is formulated for E a complex tvs. Some obvious modifications yield the proof for the real case.
To construct the required set B, we will show that $H := \bigcup_{n=1}^{\infty} n^{-1} B_n$ is absorbing and rare. If we then define $B_o := bc(H)$, we get that $B := \overline{B_o}$ is closed, balanced, absorbing and rare, thereby completing the proof.
The rareness of H can be shown in the same way as the rareness of B in the proof of 4.1.3: $V = \bigcup_{n=1}^{\infty} B_n$ since V is balanced; each B_n is rare again.
Now it would be tempting to refer to the proof of 4.1.3 again to infer $E = \bigcup_{n=1}^{\infty} nH$. This argument in itself would surely be correct, yet it does not suffice to demonstrate H to be absorbing since H need not be balanced. Neither does the passage to B_o ensure that we finally get B absorbing since nothing prevents B_o from being equal to $\{0\}$ a priori unless indeed H is absorbing.
To prove H to be absorbing, first observe that there is some $n \in \mathbf{N}$ such that $0 \in A_n$, thus $0 \in H$. Fix $0 \neq y \in E$.
From now on, we will neglect the isomorphism $\lambda \rightarrow \lambda y$ from $\mathbf{C}$ to $sp(y)$ and identify λy with λ. In order to show that H absorbs y, we will show, step by step, that some $A_p \cap sp(y)$ contains a disk $\{\alpha \in \mathbf{C} \mid \ \mid \alpha - z \mid \leq \delta\}$, that $B_n \cap sp(y)$ (for n big enough) contains the annulus

$$D := \left\{ \alpha \in \mathbf{C} \mid \ \mid z \mid - \frac{\delta}{2} \leq \mid \alpha \mid \leq \mid z \mid + \frac{\delta}{2} \right\} \qquad (*)$$

and that $H \cap sp(y)$ contains $\{\alpha \in \mathbf{C} \mid \ \mid \alpha \mid \leq \frac{|z|}{N} + \frac{\delta}{2N}\}$ for a suitable N.
The following diagram will make the remainder of the proof transparent for the reader.

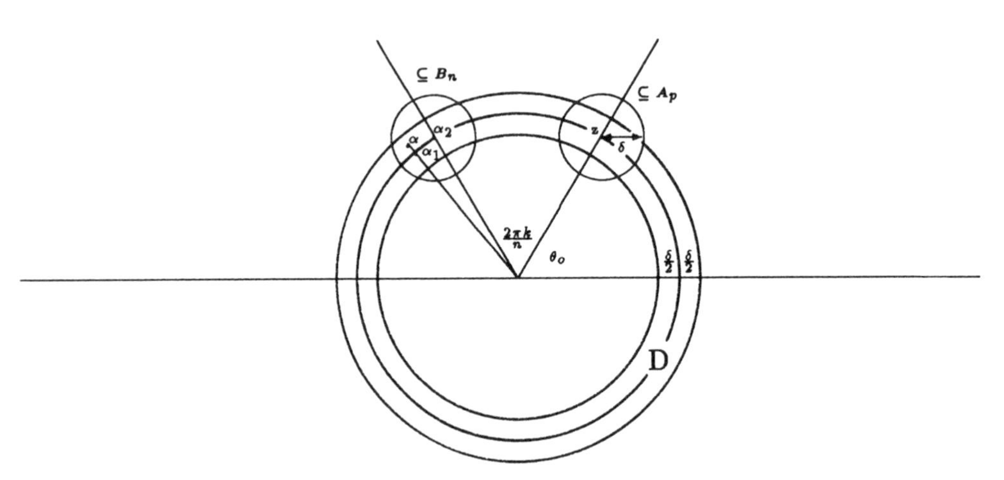

$V_1 := V \cap sp(y)$ is a closed subset of $\mathbf{C}$ either with its standard topology or the trivial topology (4.1.2) and therefore a Baire space itself. $V_1 = \bigcup_{n=1}^{\infty}(V_1 \cap A_n)$ implies that some A_p must contain a disk

$$\{\alpha \in \mathbf{C} \mid \ \mid \alpha - z \mid \leq \delta\} \quad (z = \mid z \mid e^{i\theta_o} \in \mathbf{C},\ \delta > 0).$$

B_p, by definition, contains replicas of this disk distributed along the circle with radius $\mid z \mid$. To make these replicas overlap (thereby obtaining the annulus mentioned above) we use the uniform continuity of $\theta \rightarrow \mid z \mid e^{i\theta}$ on $0 \leq \theta \leq 2\pi$ to get an integer $q \geq p$ such that

$$\mid ze^{i\theta_1} - ze^{i\theta_2} \mid \leq \frac{\delta}{2} \quad \text{for} \quad \mid \theta_1 - \theta_2 \mid \leq \frac{2\pi}{q}.$$

Now for $n \geq q$, B_n indeed contains D (defined in $(*)$):
For $\alpha = \rho e^{i\theta} \in D$ consider

$$\alpha_1 := \mid z \mid e^{i\theta} \text{ and } \alpha_2 := \mid z \mid e^{i\left(\theta_o + \frac{2\pi k}{n}\right)} = ze^{i\frac{2\pi k}{n}}$$

where $k \in \{0, ..., n-1\}$ is chosen in such a way that $\mid \theta - \theta_o - \frac{2\pi k}{n} \mid \leq \frac{2\pi}{n} \leq \frac{2\pi}{q}$. The latter condition implies $\mid \alpha_1 - \alpha_2 \mid \leq \frac{\delta}{2}$. Together with $\mid \alpha - \alpha_1 \mid = \mid \rho - \mid z \mid \mid \leq \frac{\delta}{2}$ we get $\mid \alpha - \alpha_2 \mid \leq \delta$. Thus

$$\mid e^{-i\frac{2\pi k}{n}}\alpha - z \mid = \mid e^{-i\frac{2\pi k}{n}}\alpha - e^{-i\frac{2\pi k}{n}}\alpha_2 \mid = \mid \alpha - \alpha_2 \mid \leq \delta,$$

which shows $e^{-i\frac{2\pi k}{n}}\alpha \in A_p \subseteq A_n$ and, finally, $\alpha \in B_n$.
Summarizing, B_n contains D provided $n \geq q$. Thus H contains $\bigcup_{n=q}^{\infty} \frac{1}{n}D$. If $z = 0$ then $\frac{1}{q}D$ is a full disk centered at 0, so that H absorbs y.
If $z \neq 0$, the annuli $\frac{1}{n}D$ and $\frac{1}{n+1}D$ have no space in between them, beginning from a certain value of n, say N ($N \geq q$), so, again, H - containing the disk $\{\alpha \in \mathbf{C} \mid \; |\alpha| \leq \frac{|z|}{N} + \frac{\delta}{2N}\}$ - absorbs y. N is determined by comparing the relevant radii, i.e. by solving the inequality

$$\frac{1}{N+1}\left(|z| + \frac{\delta}{2}\right) \geq \frac{1}{N}\left(|z| - \frac{\delta}{2}\right).$$

This is equivalent to $(2N+1) \geq \frac{2|z|}{\delta}$, resulting in $N \geq \frac{|z|}{\delta} - \frac{1}{2}$ $(N \geq q)$. ■

4.1.6 Remark 4.1.5 can be restated in two ways:

(i) A tvs is a Baire space if and only if it has the t-property ([57], p.224).

(ii) A tvs is Baire if and only if it is W-barrelled ([58], p.44).

Before we proceed we remind the reader of a simple fact from general topology:

4.1.7 Lemma *If B is an open subset of a topological space X and $A \subseteq X$, then*

$$\overline{A} \cap B \subseteq \overline{A \cap B}.$$

Proof. Choose $x \in \overline{A} \cap B$ and U an open neighborhood of x. Then $B \cap U$ is a neighborhood of x, so that $A \cap (B \cap U) = (A \cap B) \cap U \neq \emptyset$, implying $x \in \overline{A \cap B}$. ■

If a tvs fails to be Baire then, by 4.1.5, it can be covered by an increasing sequence of rare, balanced sets. This result is improved by

4.1.8 Theorem *A tvs E is a Baire space if and only if it cannot be covered by an increasing sequence of rare sets, each of which is closed under scalar multiplication.*

Proof. Choose x, U, V and B as in the proof of 4.1.5. For $n \in \mathbf{N}$, set

$$K_n := \{y \in B \mid ny \notin V\} \quad \text{and} \quad L_n := \{\lambda k_n \mid \lambda \in \mathbf{C}\,,\, k_n \in K_n\}.$$

If $w \in L_n \cap \frac{1}{n}V$, then $w = \alpha y$, $y \in B$ and $y \notin \frac{1}{n}V$. Hence $|\alpha| < 1$, since $\frac{1}{n}V$ is balanced. It follows that $w \in B$ since B is balanced. That is, $L_n \cap \frac{1}{n}V$ is rare because it is contained in the rare set B. Next we show that even L_n is rare:
Suppose there exists $\emptyset \neq W \subseteq \overline{L_n}$, W open. Choose $x \in W$ and $\lambda > 0$ such that

$\lambda x \in \frac{1}{n}int(V)$. $W_1 := \lambda W \cap \frac{1}{n}int(V)$ is open and contains λx. By continuity of scalar multiplication, $\overline{L_n}$ is also closed under this operation. 4.1.7 yields

$$\emptyset \neq W_1 \subseteq \overline{L_n} \cap \frac{1}{n}int(V) \subseteq \overline{L_n \cap \frac{1}{n}int(V)} \subseteq \overline{L_n \cap \frac{1}{n}V},$$

a contradiction. $L_o := \{y \in V \mid ny \in V \text{ for } n \in \mathbf{N}\}$ is closed under scalar multiplication since V is balanced. Now $\bigcup_{n=0}^{\infty} L_n$ is closed under scalar multiplication and contains the absorbing set $B \subseteq V$. Hence $E = \bigcup_{n=0}^{\infty} L_n$. Set $S_n := \bigcup_{k=0}^{n} L_k$. To see that $(S_n)_n$ satisfies the requirements of the theorem it only remains to show that L_o is rare. $L_o = \bigcap_{n=1}^{\infty} \frac{1}{n}V$, so L_o is closed.
Suppose that for some non-void open set A, $A \subseteq L_o$ and choose $x \in A$. Then $-x \in L_o$ and therefore the set $A_1 := -x + A$, which is a neighborhood of zero, is contained in $L_o + L_o$. Thus

$$E = \bigcup_{n=1}^{\infty} nA_1 \subseteq L_o + L_o \subseteq V + V \subseteq U \subseteq E \setminus \{x\},$$

a contradiction. ∎

One cannot, in general, additionally require the sets in 4.1.8 to be closed under vector addition. This is due to the fact that, in the terminology of the following chapters, there are quasi Baire spaces which are not Baire.

4.2 Inheritance Properties - the Wilansky-Klee Conjecture

In general, Baire spaces show rather poor inheritance properties:
Assuming the continuum hypothesis, J.C.OXTOBY gave an example of a completely regular Baire space whose Cartesian product with itself is not a Baire space ([38], Th.5). The example is, however, not a tvs but only a topological space. In 1982 it was shown by J.ARIAS DE REYNA that there even exist normed Baire spaces E, F such that $E \times F$ is not a Baire space (see [2]).
On the positive side we have the following result (cf. [29], p.43):
An arbitrary product of Fréchet spaces is Baire. This is a special case of a theorem by N.BOURBAKI ([6]), according to which the cartesian product of any family of complete metrizable topological spaces is a Baire space.
Furthermore, OXTOBY proved in [38], Th.3 that an arbitrary product of Baire

spaces, each of which has a countable pseudo-base, is a Baire space. Here a pseudo-base denotes a family $\mathcal{B}$ of non-empty open sets in a topological space X such that every non-empty open subset of X contains at least one member of $\mathcal{B}$. From this it can be concluded that an arbitrary product of metrizable and separable locally convex Baire spaces is Baire (cf. [5], p.30).
It is well known that any open subset of a Baire space is itself a Baire space (see [51], p.133).
For subspaces of lcs's we have

4.2.1 Theorem *Let E be a locally convex Baire space and let M be a closed, countable-codimensional subspace of E. Then M is finite-codimensional and a Baire space.*

Proof. That M must be finite-codimensional follows from 6.2.8. By finite induction, we may suppose that M is 1-codimensional, so that $E = M + sp(x)$ for some $x \in E$.
Assume that M is not Baire. Then by 4.1.5 there exists $B \subseteq M$ which is absorbing, balanced, closed and rare in M. $B' := B + \{\lambda x \mid \ \mid \lambda \mid \leq 1\}$ is closed, being the sum of a closed set and a compact set. Since furthermore B' is absorbing and balanced, it is not rare by 4.1.5. Thus B' is a neighborhood of some $\alpha x + y \in E$, where $\alpha \in \mathbf{K}$ and $y \in M$.
But then $-\alpha x + B'$ is a neighborhood of y in E, so that $(-\alpha x + B') \cap M = B$ is a neighborhood of y in M, a contradiction. ∎

For further inheritance properties of locally convex Baire spaces, see 6.3.5 and 6.3.6.

In [26], A.WILANSKY and V.KLEE conjectured that

(i) A non-zero linear functional on a Banach space E is continuous if and only if its null space can be covered by countably many sets rare in E.

This statement is equivalent to

(ii) Every dense, 1-codimensional subspace of a Banach space E is Baire.

To prove the equivalence of (i) and (ii) we need

4.2.2 Lemma *Let X be a topological space, $A \subseteq Y \subseteq X$ and Y dense in X. If A is rare in X then it is also rare in Y.*

Proof. Suppose A is not rare in Y. Then there exists some open subset $G \neq \emptyset$ of X such that $\overline{A} \cap Y \supseteq G \cap Y \neq \emptyset$. We claim that $G \subseteq \overline{A}$:
Otherwise $G \setminus \overline{A}$ would be a nonempty open subset of X, so that $Y \cap (G \setminus \overline{A}) \neq \emptyset$. But then $Y \cap G \not\subseteq Y \cap \overline{A}$, a contradiction. Hence $G \subseteq \overline{A}$ and A is not rare in X. ∎

$(i) \Rightarrow (ii)$ If M is a dense, 1-codimensional subspace of E, then $M = f^{-1}(0)$ for some discontinuous linear functional f. If $M = \bigcup_{n=1}^{\infty} A_n$, then by (i) some A_n is rare in E and therefore rare in M by 4.2.2.
$(ii) \Rightarrow (i)$ If f is continuous, then $M = f^{-1}(0)$ is a closed subspace of E. Hence M is rare in E and we can set $A_n = M$ for each n.
If M is discontinuous, then $M = f^{-1}(0)$ is dense in E. Suppose $M = \bigcup_{n=1}^{\infty} A_n$, where A_n is rare in E for each n. By 4.2.2, every A_n is rare in M, contradicting (ii). ∎

In 1980, J.ARIAS DE REYNA showed that if Martin's axiom holds, then in every separable, infinite-dimensional Banach space there exist dense hyperplanes of first category, thereby giving a negative answer to the WILANSKY-KLEE conjecture.
Martin's axiom is weaker than the continuum hypothesis and consistent with ordinary set theory (including the axiom of choice) but is independent of the continuum hypothesis.
We will see in 5.2.10 that by requiring the sets in (i) to be absolutely convex one gets a true statement.

4.3 References

4.1.1 seems to be widely known. 4.1.3 is due to S.A.SAXON and A.R.TODD ([48]). The characterizations 4.1.5 and 4.1.8 were introduced by S.A.SAXON in [47]. The WILANSKY-KLEE conjecture was first published in[26].
Finally, the result on dense hyperplanes of Banach spaces by J.ARIAS DE REYNA appeared in [1]. Further informations on Martin's axiom can also be found in [1]. For a comprehensive bibliography on permanence properties of Baire spaces see [5], 1.4.

5 Unordered Baire-like and (db)-Spaces

The characterizations of linear Baire spaces given in the previous chapter suggest several straightforward generalizations by succesively specializing the sets supposed to cover a given space.
Though at first sight this procedure might appear to be mere formal play, a closer examination will demonstrate that the spaces obtained this way serve a number of purposes. First of all, they allow a deeper understanding of Baire spaces themselves, especially concerning their inheritance properties. Moreover, many closed graph theorems do not utilize the full Baire property but only require certain spaces to be 'almost Baire' (in a sense to be made precise later on).
Finally, these 'almost Baire' spaces allow a most serviceable characterization of (LF)-spaces as we shall see in chapter 7.

5.1 Basic Concepts

5.1.1 Definition *An lcs E is called*

- convex-Baire *if E is not the union of an arbitrary sequence of rare convex sets.*
- unordered Baire-like *if E is not the union of an arbitrary sequence of rare, absolutely convex sets.*
- *a* (db)-space *(or* suprabarrelled*) if it has property (R-T-Y) : If E is covered by an increasing sequence of subspaces, then one of them is (and hence (by 5.2.3) almost all of them are) both dense and barrelled.*
- Baire like *if E is not the union of an increasing sequence of rare, absolutely convex sets.*
- quasi-Baire *if E is barrelled and is not the union of an increasing sequence of rare subspaces.*

5.1.2 Remarks

(i) Convex-Baire spaces were introduced by M.VALDIVIA in [56]. Since a thorough examination of convex-Baire spaces (including detailed proofs) is already contained in [56], chapter 1, 2., we will confine ourselves to stating the following permanence properties:
The class of all convex-Baire spaces is stable under the formation of
-separated quotients,
-completions,
-countable-codimensional subspaces and
-arbitrary products.
In fact, as we shall see in this and the following chapter, all 'almost Baire' spaces share these inheritance properties.
Moreover, VALDIVIA's book contains examples of normed convex-Baire spaces which are not Baire and of normed unordered Baire-like spaces which are not convex-Baire.

(ii) R-T-Y is an abbreviation for ROBERTSON-TWEDDLE-YEOMANS, cf.5.5.

(iii) Clearly, each locally convex Baire space is convex-Baire and each convex-Baire space is unordered Baire-like.
Every Baire-like space E is quasi-Baire.(We only have to show that E is barrelled: Suppose that B is a barrel in E which is not a neighborhood of 0. Then $E = \bigcup_{n=1}^{\infty} nB$ and each nB is rare, a contradiction.)
It will soon become clear that unordered Baire-like $\Rightarrow$ (db) $\Rightarrow$ Baire-like.

We say that an lcs E has property (R-R) (short for ROBERTSON-ROBERTSON) if it satisfies the following condition:
If E is covered by a countable collection of subspaces then one of them is both dense and barrelled.
With this notation we can formulate a generalization of a well known closed graph theorem by A.P. and W.ROBERTSON as follows:

5.1.3 Theorem *Let $(u_\iota : E_\iota \to E)_{\iota \in I}$ be a family of linear mappings from lcs's E_ι with property (R-R) into a linear space E. Equip E with the finest locally convex topology for which all the maps u_ι are continuous.*
Let $(v_n : F_n \to F)_{n \in \mathbf{N}}$ be a sequence of linear mappings from Ptàk spaces F_n into a

vector space F.

Suppose that F carries a locally convex Hausdorff topology for which each v_n is continuous. If F is covered by $(v_n(F_n))_{n\in\mathbf{N}}$, then each linear mapping $f: E \to F$ with closed graph is continuous.

Proof An inspection of the standard proof of this theorem (for Baire spaces E_ι) in [23], p.305 shows that it only requires the spaces E_ι to possess property (R-R) instead of the full Baire property. ∎

5.1.4 Theorem *For an lcs E, the following statements are equivalent:*

(i) E is unordered Baire-like.

(ii) E is not a countable union of translates of rare, absolutely convex sets.

(iii) E has property (R-R).

Proof (i)⇒(ii) Suppose $E = \bigcup_{n=1}^{\infty}(x_n + B_n)$ where each B_n is rare and absolutely convex. We show that $E = \bigcup_{m,n=1}^{\infty} mB_n$.

Otherwise there would exist some $x \in E$ such that $x \notin \bigcup_{n=1}^{\infty} sp(B_n) = \bigcup_{m,n=1}^{\infty} mB_n$. If $sp(x) \cap (x_n + B_n)$ contains two distinct points λx, μx $(\lambda \neq \mu)$, then $(\lambda - \mu)x \in B_n - B_n \subseteq sp(B_n)$, so that $x \in sp(B_n)$, a contradiction.

Thus the uncountable set $sp(x)$ meets $E = \bigcup_{n=1}^{\infty}(x_n + B_n)$ in at most countably many points, a contradiction.

Hence $E = \bigcup_{m,n=1}^{\infty} mB_n$ is not unordered Baire-like.

(ii)⇒(i) is obvious.

(i)⇒(iii) If E does not have property (R-R), then $E = \bigcup_{n=1}^{\infty} F_n$, where each F_n is rare in E or not barrelled. Choose $n \in \mathbf{N}$. If F_n is rare, define $B_n := F_n$. Otherwise let B_n be a barrel in F_n which is not a neighborhood of 0 in F_n.

In either case, B_n is a rare, absolutely convex set in E. Moreover, $E = \bigcup_{m,n=1}^{\infty} mB_n$, so that E is not unordered Baire-like.

(iii)⇒(i) Let $E = \bigcup_{n=1}^{\infty} B_n$, where each B_n is absolutely convex and define $F_n := sp(B_n)$.

If E has property (R-R) then, since $E = \bigcup_{n=1}^{\infty} F_n$, some F_n is dense and barrelled. $\overline{B_n} \cap F$ is a barrel in F_n, hence a neighborhood of 0 in F_n and therefore B_n is not rare in F_n. But F_n is dense in E, so that B_n is not rare in E by 4.2.2. Consequently, E is unordered Baire-like. ∎

5.1.4 shows that an lcs E is unordered Baire-like if and only if it is 'unordered (db)'.

Consequently, unordered Baire-like implies (db).
We claim that (db), in turn, implies Baire-like. Indeed, this follows in complete analogy to (iii)⇒(i) of 5.1.4. (The definition $F_n := sp(B_n)$ in this part of the proof conserves the property of being increasing while part (i)⇒(iii), using '$B_n := F_n$ resp. a barrel...', not necessarily does so.)
Altogether, we have thus proved the following implications:

Locally convex Baire ⇒ convex-Baire ⇒ unordered Baire-like ⇒ (db) ⇒ Baire-like ⇒ quasi-Baire ⇒ barrelled.

This and the following chapters will provide a number of examples to show that, in general, all of these concepts are distinct. For some classes of spaces, though, some of the above arrows are reversible. As a matter of fact, 1.4.15 implies the 'if'-part of

5.1.5 Proposition *A metrizable lcs E is Baire-like if and only if it is ω-barrelled. Thus for such a space the following properties are equivalent: Baire-like, quasi-Baire, barrelled, infrabarrelled and ω-barrelled.* ■

In chapter 6 we will strongly generalize 5.1.5 (cf.6.2.11).
The following result can be viewed as a dual characterization of unordered Baire-like spaces.

5.1.6 Proposition *Let E be an lcs. The following statements are equivalent:*

(i) E is unordered Baire-like.

(ii) Let $(\mathcal{A}_n)_{n\in\mathbf{N}}$ be a sequence of subsets of E' (or, more generally, of $\mathcal{L}(E,F)$, F a normed space) such that for each $x \in E$ there exists an $n \in \mathbf{N}$ so that $\{A(x) \mid A \in \mathcal{A}_n\}$ is bounded. Then for some $n \in \mathbf{N}$, $\mathcal{A}_n$ is equicontinuous.

Proof (i)⇒(ii) Let $V := \{x \in \mathbf{K} \mid \ \mid x \mid\leq 1\}$ and set $U_n := \bigcap_{A\in\mathcal{A}_n} A^{-1}(V)$ for $n \in \mathbf{N}$. For $x \in E$ there exist $n, m \in \mathbf{N}$ with $\mid A(x) \mid\leq m$ for all $A \in \mathcal{A}_n$, i.e. $x \in mU_n$.
It follows that $E = \bigcup_{m,n=1}^{\infty} mU_n$ and since each U_n is closed and absolutely convex, some U_n is a neighborhood of the origin. Consequently, $\mathcal{A}_n$ is equicontinuous. Replacing $\mathbf{K}$ by F and, accordingly, the absolute value by the norm of x yields the proof for the more general version.
(ii)⇒(i) Let $E = \bigcup_{n=1}^{\infty} V_n$, where each V_n is closed and absolutely convex and set

$\mathcal{A}_n := V_n^\circ$.
If $x \in E$ then for some $m \in \mathbf{N}$, $x \in V_m = V_m^{\circ\circ}$, implying that $\{A(x) \mid A \in \mathcal{A}_m\}$ is bounded. Therefore some $\mathcal{A}_n = V_n^\circ$ is equicontinuous, so that V_n is a neighborhood of 0. ■

5.2 Inheritance Properties

Unordered Baire-like and (db)-spaces feature a number of inheritance properties which are not known to hold for linear Baire spaces. In this section we are going to examine these properties in some detail.

5.2.1 Lemma *Let M be a closed subspace of an lcs E and H a subspace of E with $H \supseteq M$. If $\pi : E \to E/M$ is the canonical projection, then $\pi \mid_H : H \to \pi(H)$ is open.*

Proof. $H = H + M = \pi^{-1}(\pi(H))$. 3.1.5 with $G' = \pi(H)$ yields that $\pi(H) \cong H/M$ with quotient map $\pi \mid_H$. ■

5.2.2 Theorem *Let M be a closed subspace of an unordered Baire-like (resp. (db)-) space E. Then $F := E/M$ is unordered Baire-like (resp. (db)).*

Proof. Let $\pi : E \to F$ be the canonical projection and suppose $F = \bigcup_{n=1}^\infty F_n$, where each F_n is a subspace of F (and $F_n \subseteq F_{n+1}$ for each $n \in \mathbf{N}$, respectively). Then $E = \bigcup_{n=1}^\infty \pi^{-1}(F_n)$, so that for some $n_o \in \mathbf{N}$ $\overline{\pi^{-1}(F_{n_o})} = E$ and $\pi^{-1}(F_{n_o})$ is barrelled. Therefore

$$F = \pi(\overline{\pi^{-1}(F_{n_o})}) \subseteq \overline{\pi(\pi^{-1}(F_{n_o}))} = \overline{F_{n_o}}.$$

Moreover, if B is a barrel in F_{n_o} then $\pi^{-1}(B)$ is a barrel and hence a neighborhood of 0 in $\pi^{-1}(F_{n_o})$.
Thus $B = \pi(\pi^{-1}(B))$ is a neighborhood of 0 in F_{n_o} by 5.2.1 and F_{n_o} is barrelled. ■

5.2.3 Lemma *Let F be a dense and barrelled subspace of an lcs E. Then E is barrelled.*

Proof. If B is a barrel in E, then $B \cap F$ is a barrel and hence a neighborhood of 0 in F. By 1.4.8, $\overline{B \cap F}$ is a 0-neighborhood in E and so is B since it contains $\overline{B \cap F}$. ■

Notice that once B has been shown to be a 0-neighborhood in the proof of 5.2.3, 1.4.9 yields $B = \overline{B \cap F}$.

5.2.4 Theorem *If an lcs E contains a dense subspace F which is unordered Baire-like (resp. (db)), then E itself is unordered Baire-like (resp. (db)).*

Proof. If E is the union of a(n increasing) sequence of subspaces $(E_n)_n$, then $F = \bigcup_{n=1}^{\infty} F_n$, where $F_n = E_n \cap F$ for each n. By our assumption, some F_{n_o} is dense and barrelled in F.
Consequently, F_{n_o} is dense in E and contained in E_{n_o}. 5.2.3 shows that E_{n_o} is a dense, barrelled subspace of E. ∎

5.2.5 Corollary *The completion of an unordered Baire-like (resp. (db)-) space is unordered Baire-like (resp. (db)).* ∎

The discussion of permanence properties of subspaces and products requires some more preparations:

5.2.6 Proposition *If a vector space E is covered by the union of two countable families $\mathcal{F}_1, \mathcal{F}_2$ of subspaces, then one of them covers E.*

Proof. Suppose there exist $x \in E \setminus \bigcup \mathcal{F}_1$ and $y \in E \setminus \bigcup \mathcal{F}_2$. Then necessarily $x \neq y$ and the line L passing through x and y is an uncountable subset of E.
If an element F of $\mathcal{F}_k$ $(k = 1, 2)$ contains two distinct points of L then $L \subseteq F$, since F is a vector space. Thus $\{x, y\} \subseteq F$, a contradiction. It follows that $\mathcal{F}_1 \cup \mathcal{F}_2$ covers only a countable subset of the uncountable set L, which is absurd. ∎

5.2.7 Corollary *A vector space E cannot be covered by finitely many proper subspaces.*

Proof. Suppose E is covered by the proper subspaces $\{E_1, ..., E_n\}$. Set $\mathcal{F}_1 := \{E_1, ..., E_{n-1}\}$ and $\mathcal{F}_2 := \{E_n\}$.
By 5.2.6, $E \subseteq \bigcup_{k=1}^{n-1} E_k$. Continuing in this fashion we get that E is contained in some E_k, a contradiction. ∎

5.2.8 Corollary *Let F be a finite-dimensional subspace of a vector space E.*
If E is covered by a countably infinite family $\mathcal{F}$ of proper subspaces then there exists an infinite subfamily $\mathcal{F}'$ of $\mathcal{F}$ consisting of pairwise different members such that

$$F \subseteq \bigcap \{G \mid G \in \mathcal{F}'\}.$$

Proof. According to 5.2.7, $\mathcal{F} = \{F_n \mid n \in \mathbf{N}\}$, where the F_n are distinct, proper subspaces of E. Equip F with its only Hausdorff linear topology.
Then F is a Baire space and $(F_n \cap F)_n$ is a sequence of closed subspaces of F covering F. Hence there exists $n_1 \in \mathbf{N}$ such that $F \subseteq F_{n_1}$. If $n_1 < n_2 < ... < n_p$ have already been chosen with $F \subseteq F_{n_i}$ for $1 \leq i \leq p$ then by 5.2.6 (applied to $\{F_n \mid n \leq n_p\}$ and $\{F_n \mid n > n_p\}$) and by 5.2.7, E (and hence also F) is contained in $\bigcup_{n>n_p} F_n$.
The same argument as above yields the existence of $n_{p+1} > n_p$ with $F \subseteq F_{n_{p+1}}$. ■

5.2.9 Theorem *Every countable-codimensional subspace F of an unordered Baire-like space E is unordered Baire-like.*

Proof. Let $E = F + sp(\{x_n \mid n \in \mathbf{N}\})$ and $F_k := F + sp(\{x_n \mid n \leq k\})$. Then some F_k is unordered Baire-like:
Assuming the contrary would lead to $E = \bigcup_{n=1}^{\infty} F_n = \bigcup_{n,k=1}^{\infty} A_{nk}$, where the A_{nk} are absolutely convex and rare in F_n (hence in E).
Since the codimension of F in F_k is finite, it is enough to prove the theorem in the finite-codimensional case. By finite induction we may even suppose that $E = F + sp(\{x\})$ for some $x \in E$.
Assume that F is not unordered Baire-like. Then there exists a countable cover $\mathcal{G}$ of F consisting of subspaces of F none of which is both dense and barrelled. In particular, each $G \in \mathcal{G}$ is a proper subspace of F since F is barrelled (by 1.1.1).
Let $\mathcal{G}_1 := \{G \in \mathcal{G} \mid G \text{ is dense in } F\}$ and $\mathcal{G}_2 := \mathcal{G} \backslash \mathcal{G}_1$. $\mathcal{G}_1$ doesn't cover F. Otherwise $\{G + sp(x) \mid G \in \mathcal{G}_1\}$ would cover E. But then some $G + sp(x)$ and therefore some $G \in \mathcal{G}_1$ would be barrelled, a contradiction since the barrelled members of $\mathcal{G}$ belong to $\mathcal{G}_2$.
By 5.2.6, then, $\mathcal{G}_2$ covers F. Consequently, $\mathcal{H} := \{\overline{G} \mid G \in \mathcal{G}_2\}$ covers F with closed, proper subspaces of E none of which contains F. Now set

$$\mathcal{H}_o := \{H_1 \cap H_2 \mid H_1, H_2 \text{ are distinct members of } \mathcal{H}\}.$$

Applying 5.2.8 to the family $\{\overline{G} \cap F \mid G \in \mathcal{G}_2\}$ (covering F) and the one-dimensional subspace $sp(y)$ (for some $y \in F$) shows, in particular, that y is contained in two distinct spaces $(\overline{G_1} \cap F)$ and $(\overline{G_2} \cap F)$ and hence in some member of $\mathcal{H}_o$. Thus

$$F \subseteq \bigcup_{H_o \in \mathcal{H}_o} H_o.$$

If $H_o \in \mathcal{H}_o$, then its codimension in E is at least 2. It follows that $\{H_o + sp(x) \mid H_o \in \mathcal{H}_o\}$ is a countable cover of E consisting of closed proper subspaces, a contradiction.

Thus F is unordered Baire-like. ■

5.2.10 Remark 5.2.9 yields, in particular, that every dense, one-codimensional subspace of a Banach space is unordered Baire-like. Thus an argumentation similar to that in 4.2 shows that the following weakening of the WILANSKY-KLEE conjecture is a true statement:

A nonzero linear functional on a Banach space E is continuous if and only if its null space can be covered by countably many absolutely convex rare subsets of E.

5.2.11 Theorem *Every countable-codimensional subspace F of a (db)-space E is a (db)-space.*

Proof. Let $(F_n)_n$ be an increasing sequence of subspaces of F whose union is F. By 6.3.4 (proved independently), F is Baire-like so that some F_k is dense in F. Denote by G an algebraic complement of F.
$(F_n + G)_n$ is an increasing sequence of subspaces covering E. Hence there exists some $l \geq k$ such that $F_l + G$ is barrelled and dense in E. By 1.5.3 it follows that F_l is barrelled and clearly it is also dense in F. ■

In order to deal with products of unordered Baire-like spaces we need some general facts about cartesian products of tvs's. If $J \subseteq I$ are indexing sets and for $\iota \in I$ E_ι is a tvs we will canonically identify $\prod_{\iota \in J} E_\iota$ as a subspace of $\prod_{\iota \in I} E_\iota$.

5.2.12 Lemma *Let $(E_m)_m$ be a sequence of tvs's and $\mathcal{B}$ a countable family of closed, absolutely convex subsets of $E = \prod_{m \in \mathbf{N}} E_m$. If $\mathcal{B}$ covers E and*

$$\mathcal{B}_1 = \left\{ B \in \mathcal{B} \mid \prod_{m > M} E_m \subseteq B \text{ for some } M \right\},$$

then $\mathcal{F}_1 := \{sp(B) \mid B \in \mathcal{B}_1\}$ covers E.

Proof. Set $\mathcal{B}_2 := \mathcal{B} \setminus \mathcal{B}_1$ and $\mathcal{F}_2 := \{sp(B) \mid B \in \mathcal{B}_2\}$. Then

$$\bigcup (\mathcal{F}_1 \cup \mathcal{F}_2) \supseteq \bigcup \mathcal{B} = E$$

so that by 5.2.6 we only have to show that $\mathcal{F}_2$ does not cover E, i.e. that $\mathcal{A} := \{kB \mid B \in \mathcal{B}_2, k \in \mathbf{N}\}$ does not cover E. If $\mathcal{A} = \emptyset$, we are done.
Otherwise let $\mathcal{A} = \{B_n \mid n \in \mathbf{N}\}$. Then

$$\prod_{m > M} E_m \not\subseteq B_n \text{ for all } M, n \in \mathbf{N} \qquad (*)$$

As a next step we inductively prove the existence of a strictly increasing sequence $(m_k)_k$ in $\mathbf{N}_o$ and a sequence $(x_k)_k$ in E such that for each $k \geq 1$

(i) $x_k \in \prod_{m \geq m_{k-1}} E_m$ and $\sum_{i \leq k} x_i \notin B_k$, and

(ii) $\sum_{i \leq k} x_i + y \notin B_k$ for all $y \in \prod_{m \geq m_k} E_m$.

Choose $x_1 \in E \setminus B_1$ and set $m_o = 1$. Then (i) is satisfied for $k = 1$ and (ii) is trivially satisfied for $k < 1$. Suppose $m_o < m_1 < \ldots < m_{n-1}$ and $x_1, \ldots, x_n$ have already been chosen such that (i) holds for $k \leq n$ and (ii) for $k < n$.

If there were $y_i \in \prod_{m \geq i} E_m$ with $\sum_{j \leq n} x_j + y_i \in B_n$ for each i, then $\sum_{j \leq n} x_j + y_i$ would converge to $\sum_{j \leq n} x_j$. But then $\sum_{j \leq n} x_j \in B_n$ since B_n is closed, providing a contradiction to induction hypothesis (i).

Thus there exists $m_n > m_{n-1}$ with

$$\sum_{j \leq n} x_j + y \notin B_n \text{ for all } y \in \prod_{m \geq m_n} E_m,$$

so that (ii) is satisfied for all $k < n+1$. If

$$\sum_{j \leq n} x_j + \prod_{m \geq m_n} E_m \subseteq B_{n+1},$$

then $\prod_{m \geq m_n} E_m \subseteq B_{n+1}$ since B_{n+1} is absolutely convex. But this contradicts (*). Hence there exists

$$x_{n+1} \in \prod_{m \geq m_n} E_m \text{ with } \sum_{j \leq n+1} x_j \notin B_{n+1}$$

and the induction step is complete. Since $(m_k)_k$ is strictly increasing and $x_k \in \prod_{m \geq m_{k-1}} E_m$ for each k, the coordinate-wise sums $\sum_{j \geq k+1} x_j$ exist and are contained in $\prod_{m \geq m_k} E_m$. This, together with (ii), shows that

$$x := \sum_{j \leq k} x_j + \sum_{j \geq k+1} x_j \notin B_k$$

for each k. Thus

$$x \in E \setminus \bigcup_{k=1}^{\infty} B_k = E \setminus \bigcup \mathcal{F}_2.$$

■

5.2.13 Lemma *Let $(E_m)_m$ be a sequence of vector spaces and $\mathcal{F}$ a countable family of proper subspaces of $E = \prod_{m\in\mathbf{N}} E_m$. Suppose that each $F \in \mathcal{F}$ contains $\prod_{m>M} E_m$ for some $M \in \mathbf{N}$. If $\mathcal{F}$ covers E, then for some m the set $\mathcal{F}_m := \{F \in \mathcal{F} \mid E_m \not\subseteq F\}$ covers E_m.*

Proof. First we show that $\mathcal{F} = \bigcup_{m\in\mathbf{N}} \mathcal{F}_m$. Otherwise there would exist an $F \in \mathcal{F}$ such that $E_m \subseteq F$ for all m. Choose $M \in \mathbf{N}$ with $\prod_{m>M} E_m \subseteq F$. Then

$$E = \prod_{m\leq M} E_m + \prod_{m>M} E_m \subseteq F,$$

contradicting the fact that F is proper.
Now we turn to the main part of the proof. Under the assumption that no $\mathcal{F}_m$ covers E_m we will construct an element x of E not contained in any member of $\mathcal{F}$.
For each $m \in \mathbf{N}$ let

$$x_m \in E_m \setminus \bigcup \mathcal{F}_m \tag{1}$$

We inductively prove the existence of a sequence $(\alpha_m)_m$ in $\mathbf{K}$ such that

$$\sum_{i\leq m} \alpha_i x_i \notin \bigcup (\mathcal{F}_1 \cup ... \cup \mathcal{F}_m)$$

for all $m \in \mathbf{N}$. Since $x_1 \notin \bigcup \mathcal{F}_1$ we set $\alpha_1 = 1$. Suppose $\alpha_1, ..., \alpha_n$ have already been chosen with

$$\sum_{i\leq k} \alpha_i x_i \notin \bigcup (\mathcal{F}_1 \cup ... \cup \mathcal{F}_k) \text{ for } 1 \leq k \leq n \tag{2}$$

and denote by L the line through $\sum_{i\leq n} \alpha_i x_i$ and x_{n+1}. No element F of $\mathcal{F}_1 \cup ... \cup \mathcal{F}_{n+1}$ can contain two distinct points of L. Otherwise F would contain L and therefore both $\sum_{i\leq n} \alpha_i x_i$ and x_{n+1}. But this provides a contradiction to (1) or (2), according as $F \in \mathcal{F}_{n+1}$ or $F \in \bigcup(\mathcal{F}_1 \cup ... \cup \mathcal{F}_n)$. Thus since $\mathcal{G} := \bigcup(\mathcal{F}_1 \cup ... \cup \mathcal{F}_{n+1})$ is countable it covers only a countable subset of the uncountable set

$$L \setminus \{x_{n+1}\} = \{x_{n+1} + \lambda \left(\sum_{i\leq n} \alpha_i x_i - x_{n+1} \right) \mid \lambda \neq 0\}.$$

It follows that there exists $\lambda \notin \{0, 1\}$ with

$$x_{n+1} + \lambda \left(\sum_{i\leq n} \alpha_i x_i - x_{n+1} \right) \notin G := \bigcup \mathcal{G}.$$

Since G is closed under scalar multiplication, $\sum_{i \leq n} \alpha_i x_i + (\lambda^{-1} - 1)x_{n+1} \notin G$ and we can set $\alpha_{n+1} := (\lambda^{-1} - 1)$.
Define $x := \sum_m \alpha_m x_m$. Let $F \in \mathcal{F}$. To show that $x \notin F$, choose $n, M \in \mathbf{N}$ such that $F \in \mathcal{F}_n$, $M > n$ and $F \supseteq \prod_{m>M} E_m$. Put

$$y := \sum_{m \leq M} \alpha_m x_m \text{ and } z := \sum_{m>M} \alpha_m x_m.$$

Then $x = y + z$. Since $y \notin \bigcup(\mathcal{F}_1 \cup ... \cup \mathcal{F}_M)$ by (2) and $F \in \mathcal{F}_n$, it follows that $y \notin F$.
However, $z \in \prod_{m>M} E_m \subseteq F$. Consequently, $x \notin F$. ■

5.2.14 Lemma *Let I be a set and for $\iota \in I$ E_ι a tvs. Let $(B_n)_n$ be a sequence of closed subsets of $E = \prod_{\iota \in I} E_\iota$. If $\bigcup_{n \in \mathbf{N}} B_n \supseteq \prod_{\iota \in J} E_\iota$ for each countable $J \subseteq I$, then $\bigcup_{n \in \mathbf{N}} B_n = E$.*

Proof. Suppose there exists $x = (x_\iota)_{\iota \in I} \in E \setminus \bigcup_{n \in \mathbf{N}} B_n$. Since every B_n is closed, for each $n \in \mathbf{N}$ there is a finite subset J_n of I such that

$$\left((x_\iota)_{\iota \in J_n} + \prod_{\iota \in I \setminus J_n} E_\iota \right) \cap B_n = \emptyset.$$

Set $J := \bigcup_{n \in \mathbf{N}} J_n$. Then $(x_\iota)_{\iota \in J} = (x_\iota)_{\iota \in J_n} + (x_\iota)_{\iota \in J \setminus J_n} \notin B_n$ for each n although J is countable, a contradiction. ■

5.2.15 Proposition *Every countable product of unordered Baire-like spaces is unordered Baire-like.*

Proof. We only have to prove the result for a countably infinite product, since every finite product is isomorphic to an infinite one with almost all factors equal to $\{0\}$. ($\{0\}$ is obviously unordered Baire-like.)
Let $E = \prod_{n \in \mathbf{N}} E_n$ where each E_n is unordered Baire-like and suppose $\mathcal{B}$ is a countable collection of rare, absolutely convex sets covering E. By replacing each $B \in \mathcal{B}$ by its closure, we see that we can assume $\mathcal{B}$ to consist of closed, rare and absolutely convex sets.
Since E is barrelled, $sp(B)$ is a proper subspace of E for each $B \in \mathcal{B}$. Define

$$\mathcal{B}_1 := \left\{ B \in \mathcal{B} \mid \prod_{m>M} E_m \subseteq B \text{ for some } M \in \mathbf{N} \right\}.$$

By 5.2.12, $\mathcal{F}_1 := \{sp(B) \mid B \in \mathcal{B}_1\}$ covers E and by 5.2.13 there exists some $m \in \mathbf{N}$ such that $\mathcal{F}_o := \{F \in \mathcal{F}_1 \mid F \not\supseteq E_m\}$ covers E_m.
Let $\mathcal{B}_o := \{B \in \mathcal{B}_1 \mid sp(B) \not\supseteq E_m\}$. Then $\mathcal{F}_o = \{sp(B) \mid B \in \mathcal{B}_o\}$. It follows that $\{kB \mid B \in \mathcal{B}_o, k \in \mathbf{N}\}$ covers E_m. Now $(kB) \cap E_m$ is a closed, absolutely convex subset of E_m. Moreover, it is rare in E_m since its span is a proper subspace of E_m ($sp((kB) \cap E_m) = F \cap E_m \subset E_m$ since $E_m \not\subseteq F$). This contradicts the fact that E_m is unordered Baire-like and the proof is complete. ■

5.2.16 Theorem *Every arbitrary product $E = \prod_{\iota \in I} E_\iota$ of unordered Baire-like spaces E_ι is unordered Baire-like.*

Proof. Suppose there exists a countable collection $\mathcal{B}$ of closed, rare and absolutely convex sets covering E. Since E is barrelled, $\bigcup_{n \in \mathbf{N}}(nB) \subset E$ for each $B \in \mathcal{B}$.
By 5.2.14 there exists a countable subset $J_B \subseteq I$ such that $sp(B) \not\supseteq \prod_{\iota \in J_B} E_\iota$.
Set $J := \bigcup_{B \in \mathcal{B}} J_B$. Then $\{B \cap \prod_{\iota \in J} E_\iota \mid B \in \mathcal{B}\}$ is a countable cover of $F := \prod_{\iota \in J} E_\iota$ consisting of absolutely convex sets closed in F. Furthermore, each $B \cap F$ is rare in F because $sp(B \cap F) \subset F$. Since F is unordered Baire-like by 5.2.15, we arrive at a contradiction. Hence E is unordered Baire-like. ■

5.2.17 Theorem *Let E_o be the subspace of $E = \prod_{\iota \in I} E_\iota$ consisting of all $x = (x_\iota)_{\iota \in I}$ such that $x_\iota = 0$ except for at most countably many $\iota \in I$. If E is a linear Baire space, unordered Baire-like, Baire-like, quasi-Baire or barrelled, respectively, then E_o has the same property.*

Proof. Any rare subset of E_o is also rare in E. By 5.2.14, if $E_o = \bigcup_{n \in \mathbf{N}} B_n$, then $E = \bigcup_{n \in \mathbf{N}} \overline{B_n}$. ■

In order to obtain a uniform formulation of the following results we now introduce some notations: If A is a bounded, closed and absolutely convex subset of an lcs E, let E_A denote the normed space $E_A = sp(A)$ with unit ball A. If A is complete, then E_A is a Banach space (cf. [23], p.207).
Now let $E = \prod_{\iota \in I} E_\iota$ be an arbitrary product of (db)-spaces E_ι. Let E_o be as in 5.2.17 and $(F_n)_n$ an increasing sequence of subspaces of E_o whose union is E_o. For each $n \in \mathbf{N}$ let U_n be a barrel in F_n. Denote by V_n the closure of U_n in E_o and by G_n the linear span of V_n. Again, we identify each E_ι as a subspace of E.
With these definitions we have

5.2.18 Lemma *If $H = \{\iota_1, ..., \iota_r\}$ is a finite subset of I, then there exists some*

$p \in \mathbf{N}$ *such that* $\prod_{\iota \in H} E_\iota \subseteq G_n$ *for each* $n \geq p$.

Proof. Let $1 \leq k \leq r$. Since E_{ι_k} is (db) and $(F_n \cap E_{\iota_k})_{n \in \mathbf{N}}$ covers E_{ι_k}, there exists some $n_k \in \mathbf{N}$ such that $F_n \cap E_{\iota_k}$ is barrelled and dense in E_{ι_k} for each $n \geq n_k$.
Let $n \geq n_k$. Then $U_n \cap E_{\iota_k}$ is a barrel and therefore a neighborhood of 0 in $F_n \cap E_{\iota_k}$.
Now $V_n \cap E_{\iota_k} \supseteq \overline{U_n \cap E_{\iota_k}}^{E_{\iota_k}}$, so that $V_n \cap E_{\iota_k}$ is a 0-neighborhood in E_{ι_k}.
This implies that $G_n \supseteq E_{\iota_k}$ for all $n \geq n_k$. Finally, let $p := \max\{n_k \mid 1 \leq k \leq r\}$ to complete the proof. ∎

5.2.19 Remark There even exists an infinite subset P of $\mathbf{N}$ such that $E_\iota \subseteq G_p$ for all $\iota \in I$ and all $p \in P$.

Proof. Suppose the contrary is true and choose $n_1 \in \mathbf{N}$ such that, for $n \geq n_1$, no G_n contains all E_ι. Take an index $\iota_1 \in I$ with $E_{\iota_1} \not\subseteq G_{n_1}$ and assume $n_1 < ... < n_p \in \mathbf{N}$ and $\iota_1, ..., \iota_p \in I$ have already been chosen.
By 5.2.18 there exists $n_{p+1} > n_p$ with $\prod_{k=1}^{p} E_{\iota_k} \subseteq G_{n_{p+1}}$, so that we can find $\iota_{p+1} \notin \{\iota_1, ..., \iota_p\}$ with $E_{\iota_{p+1}} \not\subseteq G_{n_{p+1}}$.
For each $p \in \mathbf{N}$, let L_p be a one-dimensional subspace of E_{ι_p} not contained in G_{n_p}.
Since $\iota_1, ..., \iota_p$ are all different, $L := \prod_{p \in \mathbf{N}} L_p$ is a subspace of E_o. Moreover, L is Fréchet and hence Baire. Now $\{mV_{n_p} \mid m, p \in \mathbf{N}\}$ covers L: Observe

$$\bigcup_{m,p} mV_{n_p} \supseteq \bigcup_p G_{n_p} \supseteq \bigcup_p F_{n_p} = \bigcup_n F_n = E_o \supseteq L.$$

Thus there exists some $r \in \mathbf{N}$ such that $V_{n_r} \cap L$ is a 0-neighborhood in L. But then $G_{n_r} \supseteq L \supseteq L_r$, a contradiction. ∎

5.2.20 Proposition *Every countable product of (db)-spaces is a (db)-space.*

Proof. Analogous to the proof of 5.2.15 we may assume without loss of generality that $E = \prod_{n=1}^{\infty} E_n$ is the product of infinitely many (db)-spaces E_n.
Suppose E is not (db). Retaining the notation we agreed upon earlier, for each $m \in \mathbf{N}$ we can choose U_m in F_m such that V_m is not a neighborhood of 0 in E: No F_m is both dense and barrelled. If F_m is not dense, then $sp(V_m) = sp(\overline{U_m}) \subseteq \overline{F_m} \subset E$. If F_m is not barrelled there exists a barrel U_m in F_m which is not a neighborhood of 0 in F_m. In this case, too, V_m is not a 0-neighborhood in E.
Let $n_1 = 1$. Since E is barrelled (cf. the remarks preceding 1.1.1), there exists $x_1 \in E \setminus G_{n_1}$. Assume that $n_1 < ... < n_k \in \mathbf{N}$ and $x_1, ..., x_k \in E$ have already been constructed. By 5.2.18 there exists $n_{k+1} > n_k$ such that $\prod_{n=1}^{k} E_n \subseteq G_{n_{k+1}}$. Hence

there exists a point $x_{k+1} \in \prod_{n=k+1}^{\infty} E_n$ which is not contained in $G_{n_{k+1}}$.
Set $A := \{x_n \mid n \in \mathbf{N}\}$ and $B := \overline{\Gamma}(A)$. We claim that B is compact. Indeed, if $\pi_n : \prod_{k=1}^{\infty} E_k \to E_n$ is the canonical projection, then $\pi_n(A)$ is finite for each $n \in \mathbf{N}$. Hence $\pi_n(B) \subseteq \overline{\Gamma}(\pi_n(A))$ is relatively compact. Since $B \subseteq \prod_{n=1}^{\infty} \pi_n(B)$, our assertion is proved.
It follows that E_B is a Banach space. E and therefore also E_B are covered by the family $\{mV_{n_r} \mid m, r \in \mathbf{N}\}$ (cf. the proof of 5.2.19). Thus there exists an $r \in \mathbf{N}$ such that $V_{n_r} \cap E_B$ is a neighborhood of 0 in E_B. But then $x_r \in A \subseteq G_{n_r}$, a contradiction. ■

5.2.21 Theorem *Every arbitrary product $E = \prod_{\iota \in I} E_\iota$ of (db)-spaces E_ι is a (db)-space.*

Proof. With the notations introduced after 5.2.17 we clearly have $\overline{E_o} = E$. By 5.2.4, then, we only have to show that E_o is (db).
Under the assumption that E_o is not (db) let $(F_n)_n$ be an increasing sequence of subspaces of E_o none of which is both dense in E_o and barrelled. As in the proof of 5.2.20, for each $n \in \mathbf{N}$ we can choose U_n in F_n such that V_n is not a neighborhood of 0 in E_o.
By 5.2.17, E_o is barrelled. Hence for each $n \in \mathbf{N}$ there is a point $x_n \in E_o \setminus G_n$. Let J be a countable subset of I such that for every $n \in \mathbf{N}$ the coordinates of x_n not belonging to J vanish. Then

$$\{x_n \mid n \in \mathbf{N}\} \subseteq \prod_{\iota \in J} E_\iota =: E_J.$$

Since E_J is a (db)-space by 5.2.20, the same argument as in the proof of 5.2.18 (replace E_{ι_k} by E_J) yields that $E_J \subseteq G_k$ for some $k \in \mathbf{N}$. But then $x_k \in E_J \subseteq G_k$, a contradiction. ■

In general, neither unordered Baire-like nor (db)-spaces are stable under the formation of inductive limits. Indeed, every (LF)-space provides an example of an inductive limit of Baire spaces which is not even (db) (see 7.2.10).

5.3 Distinguishing examples

Our next aim is to demonstrate that the concepts of locally convex Baire spaces, unordered Baire-like spaces and (db)-spaces are distinct.

A short glance at the definitions shows the validity of

5.3.1 Proposition *An lcs E is unordered Baire-like if and only if the following two conditions hold*

(i) E is barrelled.

(ii) If $(A_n)_n$ is a sequence of closed, absolutely convex sets covering E, then some A_n is a barrel in E.

■

This result can be exploited as follows: If $\tau_2 \geq \tau_1$ are locally convex topologies on a vector space E such that (ii) holds for (E, τ_2), then it is also true for (E, τ_1). Moreover, every locally convex Baire space clearly satisfies (ii).
Thus if (E, τ_1) is barrelled and non-Baire and (E, τ_2) is a locally convex Baire space, then (E, τ_1) is an example of an unordered Baire-like space which is not Baire. That such spaces really do exist was first proved by W.ROBERTSON (see [39], p.255), who also gave

5.3.2 Definition *Let E be a vector space and σ, τ locally convex topologies on E. σ is called τ-*polar *if it possesses a 0-neighborhood base consisting of τ-closed sets. An lcs (E, τ) is called* ultrabarrelled *if $\sigma \leq \tau$ for each τ-polar linear topology σ on E.*

5.3.3 Remarks

(i) *Every locally convex Baire space (E, τ) is ultrabarrelled.*

Proof. Suppose σ is a τ-polar linear topology on E and let $\mathcal{U}$ be a σ-neighborhood base of 0 consisting of absolutely convex, τ-closed sets. If $U \in \mathcal{U}$, then $E = \bigcup_{n \in \mathbf{N}} nU$. Hence some nU has non-void τ-interior and therefore U is a τ-neighborhood of 0. ■

(ii) *Every ultrabarrelled space (E, τ) is barrelled.*

Proof. Denote by σ the locally convex topology on E having the τ-barrels as a neighborhood base of the origin. Then σ is τ-polar and therefore $\sigma \leq \tau$, i.e. (E, τ) is barrelled. ■

(iii) Ultrabarrelled spaces are placed *strictly* between locally convex Baire spaces and barrelled spaces. Indeed, Φ is an example of an ultrabarrelled space which is not Baire. An example of a barrelled space which is not ultrabarrelled can be found in [39], p.256.

(iv) On the other side, the concept of ultrabarrelledness is a generalization of the Baire-property (and a specialization of barrelledness) which is independent of those we studied earlier.

As a matter of fact, even in the class of all normed spaces, ultrabarrelled spaces need not be unordered Baire-like and vice versa. Concrete examples can be found in [11].

Notice that by (i) every unordered Baire-like space which is not ultrabarrelled automatically distinguishes between Baire spaces and unordered Baire-like spaces.

Now we present an example of an unordered Baire-like space which is not Baire.

5.3.4 Example Let E, F be two infinite-dimensional Banach spaces and denote by $\mathcal{F}$ the space $E' \otimes F$ equipped with the projective tensor norm (see [8], p.54). Algebraically, $\mathcal{F}$ can be identified with the space $F(E,F)$ of all bounded linear operators from E into F having finite rank.

Notice that the operator norm is dominated by the projective tensor norm ([8], p.63). Moreover, the Banach dual of the normed space $\mathcal{F}$ is isometrically isomorphic to $\mathcal{L}(F,E'')$ (the space of all bounded linear operators from F into E''), where the action of $B \in \mathcal{L}(F,E'')$ on $\sum_{i=1}^{n} f_i \otimes y_i \in E' \otimes F$ is given by $\sum_{i=1}^{n} By_i(f_i)$ ([8], p.54, 1.7).

We claim that $\mathcal{F}$ is not a Baire space: Let $\mathcal{F}_n := \{u \in \mathcal{F} \mid \text{rank}(u) \leq n\}$. $\mathcal{F}_n$ can be described as the intersection of all sets of the form

$$\left\{u \in F(E,F) \mid \det\left((f_j(u(x_i)))_{i,j=1}^{n+1}\right) = 0\right\},$$

where $x_1, ..., x_{n+1} \in E$ and $f_1, ..., f_{n+1} \in F'$. Hence every $\mathcal{F}_n$ is closed for the operator norm and consequently also for the projective tensor norm. Moreover, $\mathcal{F}_n$ is rare since the set of all operators of rank $> n$ is dense in $\mathcal{F}$. Since $\mathcal{F} = \bigcup_{n \in \mathbf{N}} \mathcal{F}_n$, our assertion is proved.

$\mathcal{F}$ is, however, unordered Baire-like. To see this, we use 5.1.6. Let $(\mathcal{B}_n)_n$ be a sequence of subsets of $\mathcal{F}' = \mathcal{L}(F,E'')$ such that, for each $F \in \mathcal{F}$, some $\mathcal{B}_n$ is bounded on F. Consider in particular $F = f \otimes y$ to obtain that for each $y \in F$, $f \in E'$ there exists some $n \in \mathbf{N}$ such that $\{\langle f \otimes y, B\rangle \mid B \in \mathcal{B}_n\} = \{By(f) \mid B \in \mathcal{B}_n\}$ is bounded. The Banach space E' is unordered Baire-like. Hence for each $y \in F$ some $\mathcal{B}_{n,y} = \{By \mid B \in \mathcal{B}_n\} \subseteq E''$ is norm bounded by 5.1.6. By the same proposition and since F is unordered Baire-like, some $\mathcal{B}_n$ is norm bounded, implying that $\mathcal{F}$ is unordered Baire-like.

Finally, we turn to the construction of examples distinguishing between unordered Baire-like and (db)-spaces.

5.3.5 Lemma *Let A be a subset of an lcs E with $sp(A) = E$. If there exists a sequence $(P_n)_n$ of continuous projections (i.e. $P_n : E \to E$, $P_n \circ P_n = P_n$) such that $P_n(A)$ is countable and $P_n(E)$ has dimension $\geq n$ for each $n \in \mathbf{N}$, then E can be covered by countably many rare subspaces.*
(It follows that E is not 'unordered quasi-Baire' (at least not unordered Baire-like, for those who dislike the notion appearing in quotation marks) and, in particular, not Baire.)

Proof.
Set $B_n := \{x \in E \mid x$ is a linear combination of strictly less than n members of $A\}$. Since $E = \bigcup_{n=2}^{\infty} B_n$, it is enough to show that each B_n can be covered by countably many closed, proper (and therefore rare) subspaces of E.
Choose $n \geq 2$ and let $A_1, A_2, ...$ be a (possibly finite) enumeration of all subsets of $P_n(A)$ with strictly less than n elements. For each $k \in \mathbf{N}$, $E_k := sp(A_k)$ is a closed subspace of E. By continuity of P_n, so is $P_n^{-1}(E_k)$.
Since $dim(E_k) < n$ and $dim(P_n(E)) \geq n$ it follows that E_k is a proper subspace of $P_n(E)$ for each k. A short computation shows that $P_n^{-1}(E_k) \cap P_n(E) = E_k$ ($P_n \circ P_n = P_n$!). Assuming $P_n^{-1}(E_k) = E$ we could derive $P_n(E) = E_k$ from this relation, which is contradictory. Thus $P_n^{-1}(E_k)$ is a closed proper subspace of E for $k \in \mathbf{N}$. Finally, $B_n \subseteq \bigcup_{k \in \mathbf{N}} P_n^{-1}(E_k)$. ■

We are now in a position to construct a wide class of metrizable (db)-spaces which are not unordered Baire-like.

5.3.6 Theorem *Every infinite-dimensional Fréchet space F contains a dense subspace which is a (db)-space but not unordered Baire-like.*

Proof. Let $0 \neq x_1 \in F$ and $f_1 \in F'$ such that $f_1(x_1) = 1$ (Hahn-Banach). If

$x_1, ..., x_n \in F$ and $f_1, ..., f_n \in F'$ have already been chosen such that $f_i(x_j) = \delta_{ij}$ $(1 \leq i, j \leq n)$, let $0 \neq x_{n+1} \in \bigcap_{i=1}^n f_i^{-1}(0)$ and let $\widetilde{f}_{n+1}$ be the linear functional on $sp(x_1, ..., x_{n+1})$ satisfying $\widetilde{f}_{n+1}(x_i) = 0$ for $1 \leq i \leq n$ and $\widetilde{f}_{n+1}(x_{n+1}) = 1$. Again by the Hahn-Banach theorem, $\widetilde{f}_{n+1}$ can continuously be extended to an $f_{n+1} \in F'$. The induction step completed we obtain a biorthogonal sequence $\{(x_n, f_n) \mid n \in \mathbf{N}\}$ in $F \times F'$. Set

$$A := \{x \in F \mid f_n(x) \in \mathbf{Q} + i\mathbf{Q} \text{ for each } n \in \mathbf{N}\}$$

and $E := sp(A)$. We claim that A (and hence E) is dense in F. Let $x \in F$ and V a closed neighborhood of 0. There exists a neighborhood U_1 of 0 with $U_1 + U_1 \subseteq V$. Beginning with exactly this U_1, construct a fundamental system $(U_n)_n$ of 0-neighborhoods such that $U_{n+1} + U_{n+1} \subseteq U_n$ for each $n \in \mathbf{N}$.
For $n \in \mathbf{N}$ choose $\epsilon_n \in \mathbf{C}$ with $\epsilon_n x_n \in U_n$ and $f_n(x + \epsilon_n x_n) = f_n(x) + \epsilon_n \in \mathbf{Q} + i\mathbf{Q}$. Then $\sum_{n=1}^\infty \epsilon_n x_n$ is a Cauchy series converging to some $y \in V$. Now

$$f_n(x + y) = f_n(x) + \sum_{m=1}^\infty \epsilon_m f_n(x_m) = f_n(x) + \epsilon_n \in \mathbf{Q} + i\mathbf{Q},$$

so that $x + y \in A \cap (x + V)$. Consequently, A is dense in F.
Define the continuous projection $P_n : F \to F$ by $P_n(x) = \sum_{i=1}^n f_i(x) x_i$ $(n = 1, 2, ...)$. A (and hence E) are invariant under P_n: If $x \in A$ then

$$f_m(P_n(x)) = f_m\left(\sum_{i=1}^n f_i(x) x_i\right) = \sum_{i=1}^n f_i(x) \delta_{im} \in \mathbf{Q} + i\mathbf{Q}$$

for all $m \in \mathbf{N}$, i.e. $P_n(x) \in A$. Hence $P_n |_E$ is a continuous projection from E to E having $x_1, ..., x_n$ in its range. Thus the hypotheses of 5.3.5 are satisfied so that E is not unordered Baire-like.
Suppose there exists an increasing sequence $(E_n)_n$ of dense, non-barrelled subspaces of E whose union is E. Again, let $(U_n)_n$ be a 0-neighborhood base consisting of absolutely convex sets U_n with $U_{n+1} + U_{n+1} \subseteq U_n$ for each $n \in \mathbf{N}$.
For $n = 1, 2, ...$ there is $B_n \subseteq F'$ such that B_n is pointwise bounded on E_n but not on F. (Assuming the contrary would lead to the contradictory result that E_n is barrelled: Consider a $\sigma(E_n', E_n)$-bounded subset G of E_n' and extend each $f \in G$ to $\widehat{f} \in F'$ by the theorem of Hahn-Banach. By our assumption, the resulting subset $\widehat{G}$ of F' is $\sigma(F', F)$-bounded and therefore equicontinuous on F because F is barrelled.

Hence G is equicontinuous on E_n and E_n is barrelled.)
In particular, the absolutely convex set B_n° is rare (otherwise B_n° would be a neighborhood of 0, contradicting the fact that B_n is not equicontinuous). Since the Fréchet space F is unordered Baire-like there exists some $x \in F \setminus \bigcup_{k,n} kB_n^\circ$, implying that each B_n is unbounded on x.
If B_1 is bounded on $x + ax_1$ for some $a \neq 0$, then B_1 is unbounded on x_1 and consequently on $x + bx_1 = (x + ax_1) + (b - a)x_1$ for all $b \neq a$. If no such a exists, then B_1 is unbounded on $x + ax_1$ for every a.
In either case there exists a_1 such that $a_1x_1 \in U_1$, $f_1(x + a_1x_1) \in \mathbf{Q} + i\mathbf{Q}$ and B_1 is unbounded on $x+a_1x_1$. There exists $g_{1,1}$ in B_1 with $| g_{1,1}(x+a_1x_1) |> 1$. Set $n_1 := 1$.
Now assume that for some $k \in \mathbf{N}$ indices $n_1, ..., n_k$, scalars $a_1, ..., a_k$ and functionals $g_{ij} \in B_{n_i}$ $(1 \leq i, j \leq k)$ have been chosen to satisfy the following conditions:

(i) $a_px_p \in U_p$ for $p \leq k$.

(ii) $f_p(x) + a_p \in \mathbf{Q} + i\mathbf{Q}$ for $p \leq k$.

(iii) B_{n_i} is unbounded on $x + a_1x_1 + ... + a_px_p$ for $1 \leq i \leq p \leq k$.

(iv) $| g_{ij}(x + a_1x_1 + ... + a_px_p) |> j$ for $1 \leq i, j \leq p \leq k$.

Then pick $n_{k+1} > n_k$ such that $E_{n_{k+1}}$ contains $\{x_1, ..., x_k\}$ $(\subseteq A \subseteq E = \bigcup_l E_l)$. $B_{n_{k+1}}$ is bounded on $x_1, ..., x_k$ and unbounded on x, hence unbounded on $x+a_1x_1+...+a_kx_k$.
Next, choose $g_{k+1,1} \in B_{n_1}, ..., g_{k+1,k} \in B_{n_k}$, $\{g_{1,k+1}, ..., g_{k+1,k+1}\} \subseteq B_{n_{k+1}}$ satisfying $| g_{ij}(x + a_1x_1 + ... + a_kx_k) |> j$.
It remains to find a suitable a_{k+1} such that (i)-(iv) are satisfied for $k + 1$ in place of k. As above, the number of scalars a such that on $x + a_1x_1 + ... + a_kx_k + ax_{k+1}$ one of the sets $B_{n_1}, ..., B_{n_{k+1}}$ is bounded, is finite. So take care that a_{k+1} is different from all these values.
Moreover, the requirements $a_kx_k \in U_k$ and $| g_{ij}(x + a_1x_1 + ... + a_{k+1}x_{k+1}) |> j$ force a_{k+1} to be sufficiently small (the set of all numbers $g_{ij}(x_{k+1})$, $i, j \leq k + 1$ is finite and therefore bounded!).
Now it is clear that we can choose a_{k+1} satisfying $f_{k+1}(x) + a_{k+1} \in \mathbf{Q} + i\mathbf{Q}$ and simultaneously being small enough to secure (i) and (iv) for $k + 1$ and avoiding the finitely many excluded values which would spoil (iii) for $k + 1$.
Thus we inductively obtain a strictly increasing sequence $(n_i)_i$ in $\mathbf{N}$, a scalar sequence $(a_i)_i$ and sequences $(g_{ij})_j$ in B_{n_i} $(i \in \mathbf{N})$ such that the Cauchy series $\sum_{i=1}^{\infty} a_ix_i$

converges to some $y \in F$, $x + y \in A$ and $| g_{ij}(x+y) | \geq j$ for all $i, j \in \mathbf{N}$.
By the definition of B_{n_i},

$$x + y \notin \bigcup_{i \in \mathbf{N}} E_{n_i} = \bigcup_{n \in \mathbf{N}} E_n = E \supseteq A,$$

a contradiction. Consequently, no such sequence $(E_n)_n$ exists. Taking $E_n = E$ for all n we see that E is barrelled. 5.1.5 now yields that E is quasi-Baire (in fact, even Baire-like).
Thus if E is the union of an increasing sequence of subspaces of E then almost all of them are dense in E and at least one of them is both dense and barrelled by what we already have proved. Hence E is a (db)-space. ■

Finally, we state without proof a similar result by M.VALDIVIA (see [55], p.576):

Every separable infinite-dimensional Fréchet space F contains a dense subspace E with the following properties:

(i) E is a (db)-space

(ii) E is not an inductive limit of unordered Baire-like spaces.

5.4 Survey

To conclude this chapter we now present a review of the concepts introduced so far.

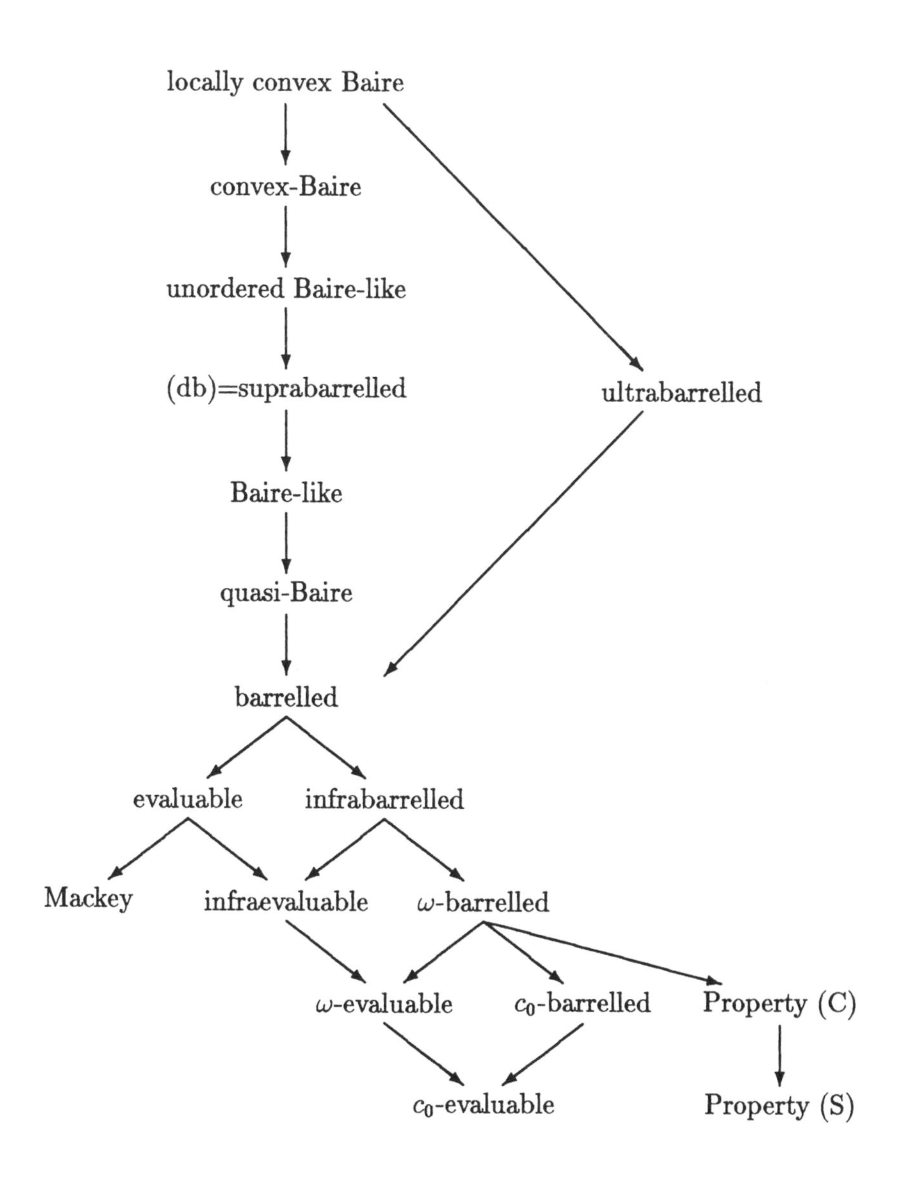

5.5 References

It was already pointed out in 5.1.2 (i) that convex-Baire spaces were introduced by M.VALDIVIA in [56].
Unordered Baire-like spaces were first considered by S.A.SAXON and A.R.TODD in [48]. (db)-spaces (or suprabarrelled spaces) were introduced independently by W.J.ROBERTSON, I.TWEDDLE and F.E.YEOMANS in [42] and by M.VALDIVIA in [55]. The concepts of Baire-like and quasi-Baire spaces are due to S.A.SAXON ([45]).
Apart from what we said in the introduction to this chapter, (db)-spaces play an important role in the study of barrelled countable enlargements (BCE'S) and barrelledly fit spaces (i.e. spaces that possess barrelled subspaces M with $codim(M) = dim(E)$):cf. [41] and [42].
5.1.3 and 5.1.4 are due to S.A.SAXON and A.R.TODD ([48]). The characterization 5.1.6 is taken from [46]. 5.2.2, 5.2.4 and 5.2.5 are due to S.A.SAXON and A.R.TODD ([48]) and M.VALDIVIA ([55]).
5.2.6 - 5.2.10 and 5.2.12 - 5.2.17 again are taken from [48]. In [55], M.VALDIVIA established the results 5.2.11 and 5.2.18 - 5.2.21.
Ultrabarrelled spaces were introduced by W.J.ROBERTSON in [39]. 5.2.4 and 5.2.5 are taken from [46], while 5.3.6 was obtained by P.P.NARAYANASWAMI and S.A.SAXON in [35].

6 Baire-like and Quasi-Baire Spaces

The main reason for a simultaneous treatment of Baire-like and quasi-Baire spaces is their close relationship to the space Φ. This, in its turn, predestines these two concepts as important tools for classifying (LF)-spaces as we shall see in chapter 7. Apart from making available some basic (inheritance) properties of Baire-like and quasi-Baire spaces we will also examine open mapping and closed graph theorems involving them.

6.1 Open Mapping and Closed Graph Theorems

6.1.1 Proposition *Let (E,τ) be a Mackey space with property (S) and $(E_n)_n$ an increasing sequence of subspaces covering E. Then E is the strict inductive limit of the sequence $((E_n, \tau\mid_{E_n}))_n$ (cf. 7.1.1).*

Proof. Let $(E,\sigma) := \varinjlim(E_n, \tau\mid_{E_n})$. Obviously, $\sigma \geq \tau$. Since τ is Mackey it suffices to show that $(E,\tau)' = (E,\sigma)'$.
Clearly, $(E,\tau)' \subseteq (E,\sigma)'$. Conversely, if $f \in (E,\sigma)'$, then $f\mid_{E_n}$ is $\tau\mid_{E_n}$-continuous for each $n \in \mathbf{N}$. By the Hahn-Banach theorem, for each $n \in \mathbf{N}$ there is a τ-continuous extension f_n of $f\mid_{E_n}$ to E. Then f is the pointwise limit of the sequence $(f_n)_n$ and since E has property (S) it follows that $f \in (E,\tau)'$. ■

6.1.2 Remarks

(i) The result is in particular valid for each barrelled space (E,τ).

(ii) Although 1.5.9 and 6.1.1 have identical conclusions, the hypotheses of both statements are not comparable as is demonstrated by 1.9.1.

6.1.3 Theorem *Let $((E_n,\tau_n))_n$ be an increasing sequence of Banach spaces such that E_n is continuously included in E_{n+1} (i.e. $E_n \hookrightarrow E_{n+1}$ is continuous) for each $n \in \mathbf{N}$.*
Endow $E = \bigcup_{n\in\mathbf{N}} E_n$ with the strongest locally convex topology τ for which the canonical injections $E_n \hookrightarrow E$ are continuous and suppose that (E,τ) is Hausdorff (i.e. an (LB)-space or a Banach space).

Let $(F_\iota)_{\iota\in I}$ be a family of Baire-like spaces and for each $\iota \in I$ let v_ι be a linear map from F_ι into a vector space F. Finally, assume that F is equipped with the strongest locally convex topology σ for which all the maps v_ι are continuous.

(i) If $g : F \to E$ is a linear map with closed graph, then g is continuous.

(ii) Each continuous linear surjection $f : E \to F$ is open.

Proof. (i) Suppose (i) has already been proved in the case that F is Baire-like and define $g_\iota : F_\iota \to E$, $g_\iota = g \circ v_\iota$ $(\iota \in I)$. Let $\varphi_\iota : F_\iota \times E \to F \times E$, $\varphi_\iota(x,y) = (v_\iota(x), y)$. Denote by G the graph of g and by G_ι the graph of g_ι $(\iota \in I)$. Then $G_\iota = \varphi_\iota^{-1}(G)$ is closed since φ_ι is continuous, so that g_ι is continuous for each $\iota \in I$. It follows that g is continuous.

Thus we can assume without loss of generality that F is Baire-like. In particular, F is barrelled. Moreover, $F = \bigcup_{n\in\mathbf{N}} g^{-1}(E_n)$ so that we only have to show the continuity of g on infinitely many of the subspaces $g^{-1}(E_n)$ (6.1.1). Since $\tau \mid_{E_n} \leq \tau_n$ it is enough to prove that $g_n := g \mid_{g^{-1}(E_n)}: g^{-1}(E_n) \to (E_n, \tau_n)$ is continuous for infinitely many $n \in \mathbf{N}$.

Let G_n and G denote the graph of g_n and g, respectively. Then $G_n = G \cap (g^{-1}(E_n) \times E_n)$. Hence G_n is closed in $(g^{-1}(E_n), \sigma \mid_{g^{-1}(E_n)}) \times (E_n, \tau \mid_{E_n})$ and therefore also in $(g^{-1}(E_n), \sigma \mid_{g^{-1}(E_n)}) \times (E_n, \tau_n)$.

Now since (E_n, τ_n) is a Ptàk space it suffices to show that infinitely many g_n : $g^{-1}(E_n) \to (E_n, \tau_n)$ are almost continuous (cf. [23], p.302, Prop.8).

Assume that g_n is not almost continuous for all $n \geq r$. Let B_r be the closed unit ball in E_r. Then $\overline{g^{-1}(B_r)} \cap g^{-1}(E_r)$ is not a neighborhood of 0 in $g^{-1}(E_r)$. Since $E_r \hookrightarrow E_{r+1}$ is continuous, B_r is bounded in E_{r+1} and therefore contained in some multiple B_{r+1} of the unit ball in E_{r+1}. Furthermore, $\overline{g^{-1}(B_{r+1})} \cap g^{-1}(E_{r+1})$ is not a neighborhood of 0 in $g^{-1}(E_{r+1})$. By induction we obtain an increasing sequence $(S_n)_{n\geq r}$ such that S_n is a multiple of the unit ball in E_n and $\overline{g^{-1}(S_n)}$ is not a neighborhood of 0 in F for each $n \geq r$.

But then $(n\overline{g^{-1}(S_n)})_{n\geq r}$ is an increasing sequence of rare, absolutely convex sets covering the Baire-like space F, a contradiction. Hence g is continuous.

(ii) Let M be a closed subspace of E and $\pi : E \to E/M$ the canonical epimorphism. Denote by $h_n : E_n/(E_n \cap M) \hookrightarrow E/M$ the mapping $x + (E_n \cap M) \to x + M$ and by $\pi_n : E_n \to E_n/(E_n \cap M)$ the canonical projection. We claim that the quotient topology ξ on E/M is identical with the finest locally convex topology η for which

all h_n are continuous.
Since $h_n \circ \pi_n = \pi \mid_{E_n}$ is continuous for each n it follows that $\eta \geq \xi$. Conversely, it is enough to prove that $\pi : E \to (E/M, \eta)$ is continuous. By the definition of the topology on E this is the case if $\pi \mid_{E_n}: E_n \to (E/M, \eta)$ is continuous for each n. Now $\pi \mid_{E_n} = h_n \circ \pi_n$ and our assertion is proved.
Next we show that, for each $n \in \mathbf{N}$, the mapping

$$\varphi_n : E_n/(E_n \cap M) \hookrightarrow E_{n+1}/(E_{n+1} \cap M),\ x + (E_n \cap M) \to x + (E_{n+1} \cap M)$$

is continuous. To this end one only has to observe that $\varphi_n \circ \pi_n = \pi_{n+1} \mid_{E_n}$.
To summarize, it follows that E/M is the same type of space as E. Now set $M := f^{-1}(0)$ and define $\overline{f} : E/M \to F$ to be the injection associated with f. To see that $\overline{f}$ (and hence f) is open one has to prove that $\overline{f}^{-1}$ is continuous. But since the graph of $\overline{f}^{-1}$ is closed in $F \times E/M$, this follows from (i). ■

6.1.4 Theorem *Let E be as in 6.1.3 and f a continuous linear map from a Baire-like space F into E. If E is Hausdorff, then $f(F) \subseteq E_n$ for some n and f is continuous as a mapping from F into the Banach space E_n.*

Proof. Since E is Hausdorff and f is continuous, the graph of f is closed in $F \times E$. Besides, the Baire-like space F is covered by the increasing sequence $(f^{-1}(E_n))_n$ of subspaces. Hence there exists some $p \in \mathbf{N}$ such that $\overline{f^{-1}(E_n)} = F$ for all $n \geq p$.
By the proof of 6.1.3 (i), for some $n \geq p$ f maps $D := f^{-1}(E_n)$ continuously into the Banach space E_n. Of course, $f \mid_D$ has a continuous linear extension $g : F \to (E_n, \tau_n)$. g satisfies the claims being made about f. We conclude the proof by showing $f = g$ (and thereby $D = F$).
Let $x \in F$. Then $x = \lim x_\alpha$ for some net $(x_\alpha)_\alpha$ in D which implies $g(x) = \lim f(x_\alpha)$ in (E_n, τ_n). On the other hand, again by the proof of 6.1.3 (i), there exists $m \geq n$ such that $x \in G := f^{-1}(E_m)$ and $f \mid_G: G \to (E_m, \tau_m)$ is continuous.
Consequently, $f(x) = \lim f(x_\alpha)$ in (E_m, τ_m). The fact that (E_n, τ_n) is continuously included in (E_m, τ_m) allows us to conclude $f(x) = g(x)$. ■

6.1.5 Remark 6.1.4 is no longer valid if the spaces $(E_n)_n$ are assumed to be only Fréchet spaces (cf. 7.2.5). It remains true, though, for Fréchet spaces E_n if F, too, is assumed to be a Fréchet space (cf. 7.1.14).

6.1.6 Proposition *Let E, F be lcs's and $A : D[A] \subseteq E \to F$ a linear map. Set $N[A] := \{x \in D[A] \mid A(x) = 0\}$ and $G[A] := \{(x, y) \in E \times F \mid x \in D[A],\ y = A(x)\}$.*

If $G[A]$ is closed, then $N[A]$ is also closed.

Proof. $j : E \to E \times F$, $x \to (x, 0)$ is continuous and

$$N[A] = \{x \in D[A] \mid (x,0) \in G[A]\} = \{x \in E \mid (x,0) \in G[A]\} = j^{-1}(G[A]).$$

■

6.1.7 Proposition *With E, F and A as in 6.1.6, consider the following diagram*

$$\begin{array}{ccc} D[A] \subseteq E & \xrightarrow{A} & F \\ \pi \downarrow & \nearrow \widehat{A} & \\ E/N[A] & & \end{array}$$

Here π denotes the canonical projection and $\widehat{A}$ is the injection associated with A. Then $G[A]$ is closed if and only if $G[\widehat{A}]$ is closed.

Proof. Let $id : E \to E$ be the identical mapping. Then

$$(\pi \times id)^{-1}(G[\widehat{A}]) = \{(x,y) \in E \times F \mid (\pi(x), y) \in G[\widehat{A}]\} =$$

$$= \{(x,y) \in E \times F \mid (\pi(x) \in \pi(D[A])) \wedge (y = \widehat{A}(\pi(x)))\} =$$

$$= \{(x,y) \in E \times F \mid (x \in D[A] + N[A]) \wedge (y = \widehat{A}(\pi(x)))\} =$$

$$= \{(x,y) \in E \times F \mid (x \in D[A]) \wedge (y = A(x))\} = G[A]$$

Hence the result follows from the definition of the quotient topology. ■

6.1.8 Theorem *Let E be a Ptàk space, F a barrelled space and A a closed linear map from $D[A] \subseteq E$ into F. If the range $R[A]$ of A is of countable codimension in F, then A is open and $R[A]$ is closed in F. If F is quasi-Baire, then the codimension of $R[A]$ in F is finite.*

Proof. Let $N[A]$ be the null space of A and denote by

$$\widehat{A} : D[\widehat{A}] = D[A]/N[A] \subseteq E/N[A] \to F$$

the injection associated with A. Then $R[\widehat{A}] = R[A]$ and $\widehat{A}$ is closed by 6.1.7. Since $N[A]$ is closed by 6.1.6, $E/N[A]$ is a Ptàk space (cf. [50], IV.8.3, Cor.3). $R[A]$ is barrelled by 1.5.3. Hence the closed map $\widehat{A}^{-1}$ (defined on $R[A]$) is continuous by a closed graph theorem by ROBERTSON/ROBERTSON (see [23], p.301, Theorem 4). Consequently, $\widehat{A}$ and therefore A is open.

Now suppose $(\widehat{A}(\widehat{x}_\iota))_\iota$, $\widehat{x}_\iota \in D[\widehat{A}]$ is a Cauchy net in $R[A]$ converging to $y \in F$. Since $\widehat{A}^{-1}$ is continuous, $(\widehat{x}_\iota)_\iota$ is a Cauchy net in $E/N[A]$. The Ptàk space $E/N[A]$ is complete by [50], IV.8.1, so that $(\widehat{x}_\iota)_\iota$ converges to some $\widehat{x}_o \in E/N[A]$. Due to the fact that $\widehat{A}^{-1}$ is closed we conclude that $\widehat{A}^{-1}(y) = \widehat{x}_o$. Thus $y = \widehat{A}(\widehat{x}_o) \in R[\widehat{A}] = R[A]$ and $R[A]$ is closed.
The last assertion will follow immedeately from 6.2.8, $(iii) \Rightarrow (i)$. ■

6.2 Interrelation with Φ

The examination of the relationship between Baire-like spaces and the space Φ requires some technical preparations. Let us remark here that Φ is barrelled but not quasi-Baire (cf. 2.1).

6.2.1 Lemma *Let $(A_n)_n$ be an increasing sequence of closed, absolutely convex subsets of an lcs E such that A_{n+1} is not contained in $sp(A_n)$ for $n \in \mathbf{N}$. Choose elements $x_n \in A_{n+1} \setminus sp(A_n)$ and set $S_n := sp(x_1, ..., x_n)$ and $S := sp(x_1, x_2, ...)$. Let p be any seminorm on S. Then there exists a sequence $(f_n)_n$ in E' such that with $q_n(x) := \max_{1 \leq r \leq n} \mid f_r(x) \mid$ the following two conditions hold for each n:*

(i) $f_n \in A_n^\circ$.

(ii) If $x \in S_n$, then $q_n(x) \geq (1 + 2^{-n})p(x)$.

Proof. S_n and S remain unchanged if we replace each x_n by a non-zero scalar multiple. Thus we can assume $p(x_n) \leq 1$. Moreover, for $x = \sum_{i=1}^n a_i x_i \in S_n$ we have $p(x) \leq \sum_{i=1}^n \mid a_i \mid$, so it is sufficient to establish (ii) for the norm $\widetilde{p}(\sum_{i=1}^n a_i x_i) = \sum_{i=1}^n \mid a_i \mid$. We write p instead of $\widetilde{p}$.
Since $x_1 \notin (1 + 2^{-1})A_1$, by the Hahn-Banach theorem there exists $f_1 \in A_1^\circ$ with $f_1(x_1) = (1 + 2^{-1})$, so that (i) and (ii) are satisfied for $n = 1$.
Assume that $f_1, ..., f_k \in E'$ have already been chosen such that (i) and (ii) hold for $n \leq k$. Since

$$1 + 2^{-k-1} < 1 + 2^{-k} = \lim_{M \to \infty} \frac{(1 + 2^{-k})M - q_k(x_{k+1})}{M + 1},$$

there exists $M_o > 0$ such that, for $M \geq M_o$,

$$1 + 2^{-k-1} \leq \frac{(1 + 2^{-k})M - q_k(x_{k+1})}{M + 1}.$$

Let $N = \max(1, (2 + 2^{-k})M_o - q_k(x_{k+1}))$. Then $N > 0$ and

$$1 + 2^{-k-1} \leq \frac{N - M_o}{M_o + 1}.$$

Due to the fact that $x_{k+1} \notin NA_{k+1}$, the Hahn-Banach theorem yields the existence of $f_{k+1} \in A_{k+1}^{\circ}$ with $f_{k+1}(x_{k+1}) = N$. For $x = \sum_{i=1}^{k} a_i x_i \in S_k$ let $y := x + x_{k+1}$ (observe $p(x_{k+1}) = 1$). Then since $x_1, ..., x_k \in A_{k+1}$ we have $| f_{k+1}(x) | \leq \sum_{i=1}^{k} | a_i | = p(x)$. Now there are two possibilities:
If $p(x) \geq M_o$, then

$$1 + 2^{-k-1} \leq \frac{(1 + 2^{-k})p(x) - q_k(x_{k+1})}{p(x) + 1} \leq \frac{q_k(x) - q_k(x_{k+1})}{p(y)} \leq \frac{q_k(y)}{p(y)} \leq \frac{q_{k+1}(y)}{p(y)}.$$

Otherwise, if $p(x) < M_o$, then

$$1 + 2^{-k-1} \leq \frac{N - M_o}{M_o + 1} \leq \frac{| f_{k+1}(x_{k+1}) | - p(x)}{p(y)} \leq$$

$$\leq \frac{| f_{k+1}(x_{k+1}) | - | f_{k+1}(x) |}{p(y)} \leq \frac{| f_{k+1}(y) |}{p(y)} \leq \frac{q_{k+1}(y)}{p(y)}.$$

In either case, $q_{k+1}(y) \geq p(y)(1 + 2^{-k-1})$. Therefore

$$q_{k+1}\left(\sum_{i=1}^{k+1} a_i x_i\right) = | a_{k+1} | \, q_{k+1}\left(\sum_{i=1}^{k} \frac{a_i}{a_{k+1}} x_i + x_{k+1}\right) \geq$$

$$\geq | a_{k+1} | \, p\left(\sum_{i=1}^{k} \frac{a_i}{a_{k+1}} x_i + x_{k+1}\right)(1 + 2^{-k-1}) = p\left(\sum_{i=1}^{k+1} a_i x_i\right)(1 + 2^{-k-1}),$$

completing the induction step. ∎

6.2.2 Corollary *Under the hypotheses of 6.2.1, if E is ω-barrelled and $(A_n)_n$ is absorbent, then S is isomorphic to Φ.*

Proof. For $x \in E$ there exist $m, n \in \mathbf{N}$ such that $x \in mA_n$. Since $(A_n)_n$ is increasing it follows that $f_{n+k} \in A_{n+k}^{\circ} \subseteq A_k^{\circ}$ for all $k \in \mathbf{N}$. Hence $| f_{n+k}(x) | \leq m$ for every $k \in \mathbf{N}$, so that $(f_n)_n$ is $\sigma(E', E)$-bounded. Thus $(f_n)_n$ is equicontinuous because E is ω-barrelled.
Let U be a neighborhood of 0 such that $| f_k(x) | \leq 1$ for all $x \in U$ and all $k \in \mathbf{N}$. Define the seminorm q on E by $q(x) = \sup q_n(x) = \sup | f_k(x) |$. Then $| q(x) | \leq 1$ for all $x \in U$ so that q is continuous.

By 6.2.1 (ii), $p \leq q \mid_S$, implying that p is continuous on S. Since p was arbitrary, S carries the strongest locally convex topology and $dim(S) = \aleph_0$, i.e. $S \cong \Phi$. ■

6.2.3 Theorem *Every barrelled space E which does not contain Φ is Baire-like.*

Proof. If E is barrelled, but not Baire-like, there exists an increasing sequence $(B_n)_n$ of rare, absolutely convex sets covering E.
Let $n_1 = 1$. Since $\overline{B_1}$ is a barrel in its span, there exists some $x_1 \in E \setminus sp(\overline{B_1})$. Choose $n_2 > n_1$ with $x_1 \in \overline{B_{n_2}}$. Now suppose $n_1, ..., n_k$ have been defined such that $x_i \in \overline{B_{n_{i+1}}} \setminus sp(\overline{B_{n_i}})$ for $i = 1, ..., k-1$. Again, since $\overline{B_{n_k}}$ is rare and E is barrelled there exists $x_k \in E \setminus sp(\overline{B_{n_k}})$ and $n_{k+1} > n_k$ with $x_k \in \overline{B_{n_{k+1}}}$.
Set $A_k := \overline{B_{n_k}}$. Then $(A_n)_n$, $(x_n)_n$ and E satisfy the hypothesis of 6.2.2, so that S is isomorphic to Φ. ■

The hypothesis of 6.2.3 cannot be weakened to E being ω-barrelled, as is demonstrated by

6.2.4 Example *An ω-barrelled space which does not contain Φ without being Baire-like.*
Let (E, τ) be a reflexive Banach space whose strong dual $(E', \beta(E', E))$ is not separable (e.g. $l^p(D)$, $p > 1$, $\mid D \mid > \aleph_0$). Let $\tau_1 = \nu(E, E')$ (cf. 1.6). Since E is barrelled, $\tau = \beta(E, E')$ by 1.6.5. From the second diagram in 1.6 it is now obvious that $\sigma(E, E') \leq \nu(E, E') \leq \tau$. 1.6.6 implies that $(E, \nu(E, E')) = (E, \tau_1)$ is ω-barrelled.
Since Φ is non-metrizable, (E, τ) and consequently (E, τ_1) do not contain Φ. We claim that (E, τ_1) is not a Mackey space (then (E, τ_1) is not Baire-like, in fact, not even evaluable). It suffices to show that $\tau_1 < \tau$.
Suppose $\tau_1 = \tau$. Then the polar of some countable subset C of E' is contained in the unit ball B of (E, τ). Therefore $C^{\circ\circ} \supseteq B^\circ$ and B° is the unit ball of the normed space E'. Now

$$C^{\circ\circ} = \overline{\Gamma}^{\sigma(E',E)} C = \overline{\Gamma}^{\sigma(E',E'')} C = \overline{\Gamma}^{\beta(E',E)} C, \text{ so that } B^\circ \subseteq C^{\circ\circ} \subseteq \overline{sp(C)}^{\beta(E',E)}.$$

Hence $sp(C)$ is dense in $(E', \beta(E', E))$. But then the rational linear combinations of elements of C are also dense in E' and E' is separable, a contradiction.
Notice that (E, τ_1) is another example of an ω-barrelled space which is not Mackey (cf. 1.9.3).

6.2.5 Corollary *If E is an lcs with property (S) and $(A_n)_n$ is an absorbent sequence* *of closed subsets of E then each bounded subset is contained in some $sp(A_n)$.*

Proof. Suppose there exists a bounded subset B of E such that $B \not\subseteq sp(A_n)$ for $n = 1, 2, \ldots$. We can and will assume B to be balanced.
Since $(A_n)_n$ is absorbent and B is balanced we can inductively define a strictly increasing sequence $(n_k)_k$ in $\mathbf{N}$ and a sequence $(x_k)_k$ in B such that $x_k \in A_{n_{k+1}} \setminus sp(A_{n_k})$ for each k. By the Hahn-Banach theorem we now inductively choose a sequence $(f_k)_k$ in E' such that $f_k \in 2^{-k} A^{\circ}_{n_k}$ and

$$| f_{k+1}(x_{k+1}) | \geq (k+1) + \sum_{j=1}^{k} | f_j(x_{k+1}) |$$

for each $k \in \mathbf{N}$. For $x \in E$ there exist $m, k \in \mathbf{N}$ with $x \in mA_{n_k}$. Now $f_{k+l} \in 2^{-(k+l)} A^{\circ}_{n_{k+l}} \subseteq 2^{-(k+l)} A^{\circ}_{n_k}$ and therefore $| f_{k+l}(x) | \leq 2^{-(k+l)} m$ for every $l \in \mathbf{N}_o$. This shows that $\sum_{k=1}^{n} f_k$ converges point-wise to some $f \in E'$ (E has property (S)).
On the other hand,

$$| f(x_{k+1}) | \geq \left(| f_{k+1}(x_{k+1}) | - \sum_{j=1}^{k} | f_j(x_{k+1}) | \right) - \sum_{j=k+2}^{\infty} | f_j(x_{k+1}) | \geq$$

$$\geq (k+1) - \sum_{j=k+2}^{\infty} 2^{-j} > k.$$

(Observe $x_{k+1} \in A_{n_{k+2}}$ which implies $| f_{k+2+l}(x_{k+1}) | \leq 2^{-(k+2+l)}$ for $l \geq 0$.) Thus $(x_k)_k$ is not $\sigma(E, E')$-bounded and therefore not bounded, contradicting the fact that $\{x_k \mid k \in \mathbf{N}\} \subseteq B$. ■

6.2.6 Corollary *Every metrizable barrelled space E is Baire-like.*

Proof. E does not contain Φ since Φ is not metrizable. ■

6.2.6 is in fact a special case of 1.4.15. That 6.2.3 is indeed a stronger result can be seen from

6.2.7 Example Let I be any indexing set with $| I | > \aleph_0$. Then $E := \mathbf{K}^I$ is barrelled and non-metrizable. By 2.3.3, E is not Φ-productive and in particular does not contain Φ. Hence E is Baire-like by 6.2.3.

6.2.8 Theorem *For a barrelled space E, the following statements are equivalent:*

(i) E is not quasi-Baire.

(ii) E contains a complemented copy of Φ.

(iii) E contains a closed, $\aleph_0$*-codimensional subspace.*

(iv) E is isomorphic to $E \times \Phi$.

(v) E is the strict inductive limit (cf. 7.1.1) of a strictly increasing sequence of closed barrelled subspaces of E.

Proof. $(ii) \Leftrightarrow (iii)$ by 2.2.1.
$(iv) \Rightarrow (ii)$ Clear.
$(ii) \Rightarrow (iv)$ For some closed subspace F of E we have

$$E \cong F \times \Phi \cong F \times (\Phi \times \Phi) \cong (F \times \Phi) \times \Phi \cong E \times \Phi.$$

$(iii) \Rightarrow (v)$ Let E_o be a closed, $\aleph_0$-codimensional subspace of E. Let $(E_n)_{n\in\mathbf{N}}$ be an increasing sequence of subspaces of E such that $E = \bigcup_{n=0}^{\infty} E_n$ and such that E_{n-1} has codimension 1 in E_n for each $n \in \mathbf{N}$.
Each E_n is obviously closed and in addition to this it is barrelled by 1.5.3. 6.1.1 yields that E is the strict inductive limit of the sequence $(E_n)_n$.
$(v) \Rightarrow (i)$ Clear, since E is the union of countably many closed, proper and therefore rare subspaces.
$(i) \Rightarrow (iii)$ Let $(E_n)_n$ be a strictly increasing sequence of closed subspaces covering E. Choose $x_1 \in E_2 \setminus E_1$. By the Hahn-Banach theorem we can find $f_1 \in E'$ with $f_1(x_1) = 1$ and $f_1 |_{E_1} = 0$.
Suppose $x_1, ..., x_k$ and $f_1, ..., f_k \in E'$ have already been defined such that $f_i(x_j) = \delta_{ij}$ $(1 \leq i, j \leq k)$, $x_i \in E_{i+1} \setminus E_i$ and $f_i |_{E_i} = 0$ for $1 \leq i \leq k$. Let $y \in E_{k+2} \setminus E_{k+1}$ and set $x_{k+1} = y - \sum_{i=1}^{k} f_i(y)x_i$. Then $x_{k+1} \in E_{k+2} \setminus E_{k+1}$ and we can choose $f_{k+1} \in E'$ such that $f_{k+1}(x_{k+1}) = 1$ and $f_{k+1} |_{E_{k+1}} = 0$.
Continuing in this fashion we obtain a biorthogonal sequence $(x_i, f_i)_{i\in\mathbf{N}}$ with $x_i \in E_{i+1} \setminus E_i$ and $f_i |_{E_i} = 0$ for each $i \in \mathbf{N}$. $E_o := \bigcap_{i=1}^{\infty} f_i^{-1}(0)$ is then a closed subspace of E and $E_o \cap sp(\{x_i \mid i \in \mathbf{N}\}) = \{0\}$.
Finally, we show that the codimension of E_o in E is $\aleph_0$. Let $x \in E$ and choose $n \in \mathbf{N}$ such that $x \in E_{n+1}$. Then $f_j(x) = 0$ for all $j \geq n-1$. Setting $z = \sum_{i=1}^{n} f_i(x)x_i$ it follows that $x - z \in E_o$, $z \in sp(\{x_i \mid i \in \mathbf{N}\})$ and $x = (x - z) + z$. Hence $sp(\{x_i \mid i \in \mathbf{N}\})$ is an algebraic complement of E_o in E. ■

6.2.9 Corollary *If an lcs E does not contain* Φ *complemented, then it is barrelled* *if and only if it is quasi-Baire.* ■

6.2.10 Theorem *If a metrizable lcs E has property (S), then it is Baire-like.*

Proof. Because of 6.2.6 it is enough to show that E is barrelled.
Suppose there exists a closed subset $B \subseteq E'$ which is $\sigma(E', E)$-bounded but not equicontinuous. Let $(U_n)_n$ be a 0-neighborhood base such that $U_{n+1} \subseteq U_n$ for each $n \in \mathbf{N}$. Then B is not uniformly bounded on U_n $(n = 1, 2, ...)$.
We inductively construct a subsequence $(U_{n_k})_k$ of $(U_n)_n$, a sequence $(f_k)_k$ in B and sequences $(M_k)_k$ of constants and $(x_k)_k$ in E such that the following conditions hold:

(i) $M_k \geq 1 \; \forall k \in \mathbf{N}$.

(ii) $x_k \in U_{n_k} \; \forall k \in \mathbf{N}$.

(iii) $| f_p(x_k) | \leq 1$ for $p < k$.

(iv) $| f_k(x_k) | \geq k M_k 2^k \; \forall k \in \mathbf{N}$.

(v) $| f_p(x_k) | \leq M_p$ for $p > k$.

Set $M_1 = 1$, $n_1 = 1$ and choose $x_1 \in U_{n_1}$ and $f_1 \in B$ with $| f_1(x_1) | \geq M_1 2^1$. Then (i)-(v) are satisfied for $p, k \leq 1$.
Assume that $n_1 < ... < n_r \in \mathbf{N}$, $f_1, ..., f_r \in B$, $M_1, ..., M_r$ and $x_1, ..., x_r$ have been chosen to satisfy (i)-(v) for all $p, k \leq r$. Let $n_{r+1} > n_r$ such that $| f_p(x) | \leq 1$ for all $x \in U_{n_{r+1}}$ and $1 \leq p \leq r$. Since B is $\sigma(E', E)$-bounded there exists $M_{r+1} \geq 1$ with $| f(x_k) | \leq M_{r+1}$ for all $f \in B$ and $1 \leq k \leq r$.
Moreover, since B is not uniformly bounded on $U_{n_{r+1}}$ we can find $f_{r+1} \in B$ and $x_{r+1} \in U_{n_{r+1}}$ such that $| f_{r+1}(x_{r+1}) | \geq (r+1) M_{r+1} 2^{r+1}$. Hence (i)-(v) are valid for $p, k \leq r + 1$ and the induction step is completed.
Now $\sum_{k=1}^{\infty} (2^k M_k)^{-1} f_k(x)$ converges absolutely for each $x \in E$ because $(f_k)_k \subseteq B$ is pointwise bounded and $M_k \geq 1$ for each k. Since E has property (S),

$$f(x) := \sum_{p=1}^{\infty} (2^p M_p)^{-1} f_p(x)$$

defines a continuous linear functional on E. For each $k \in \mathbf{N}$ we have

$$| f(x_k) | \geq (2^k M_k)^{-1} | f_k(x_k) | - \sum_{p \neq k} (2^p M_p)^{-1} | f_p(x_k) | \geq$$

$$\geq (2^k M_k)^{-1} k M_k 2^k - \sum_{p=1}^{k-1}[(2^p M_p)^{-1}] - \sum_{p=k+1}^{\infty}[(2^p M_p)^{-1} M_p] \geq k - \sum_{p=1}^{\infty} 2^{-p} = k-1.$$

It follows that $| f(x_k) | \to \infty$ although $x_k \to 0$ and f is continuous, a contradiction. Therefore no such B exists and E is barrelled. ■

6.2.10, together with the implications we have proved thus far (cf. 5.4) yields

6.2.11 Corollary *For a metrizable lcs E, the following properties are equivalent: Baire-like, quasi-Baire, barrelled, infrabarrelled, ω-barrelled, property (C) and property (S).* ■

6.2.12 Theorem *If an lcs E contains a dense, Baire-like subspace M, then E itself is Baire-like.*

Proof. Suppose E is the union of an increasing sequence $(A_n)_n$ of closed, absolutely convex sets. Then $M = \bigcup_{n \in \mathbf{N}}(A_n \cap M)$, so that some $A_n \cap M$ is a neighborhood of 0 in M. By 1.4.8 (ii), $\overline{A_n \cap M}$ is a neighborhood of 0 in E. Hence $A_n = \overline{A_n} \supseteq \overline{A_n \cap M}$ is a neighborhood of the origin in E, implying that E is Baire-like. ■

6.2.13 Corollary *The completion of a Baire-like space is Baire-like.* ■

The proof of 1.4.15, together with 6.2.12 yields

6.2.14 Theorem *An ω-barrelled space E is Baire-like if and only if its completion is Baire-like.* ■

6.2.15 Corollary *In the variety $\mathcal{V}^w$ of all lcs's having their weak topology, the following concepts coincide:*
Baire-like, quasi-Baire, barrelled, infrabarrelled and ω-barrelled.

Proof. If $E \in \mathcal{V}^w$ then by the proof of 2.3.3, (4)$\Rightarrow$(1), E is isomorphic to a dense subspace of a product of copies of $\mathbf{K}$. Thus the completion of E is isomorphic to a product of copies of $\mathbf{K}$ and therefore is a Baire space (in particular, Baire-like). ■

6.3 Inheritance Properties

6.3.1 Theorem *Let M be a dense, ω-barrelled subspace of an lcs E. M is quasi-Baire if and only if E is quasi-Baire.*

Proof. If M is quasi-Baire then, in particular, it is barrelled and therefore E is barrelled by 5.2.3. Let $(E_n)_n$ be an increasing sequence of subspaces covering E. Since $M = \bigcup_{n\in\mathbf{N}}(M \cap E_n)$, some $M \cap E_n$ is not rare in M and therefore dense in M. Thus $E = \overline{M} \subseteq \overline{E_n}$ and E is quasi-Baire.

Conversely, let E be quasi-Baire. First we show that M is barrelled. Let $(A_n)_n$ be an increasing sequence of absolutely convex sets whose union is M. Assume that the subspace $F := \bigcup_{n\in\mathbf{N}} n\overline{A_n}$ is proper and choose $x_o \in E \setminus F$.

Since $\overline{A_n} = \overline{A_n}^{\circ\circ}$, for each $n \in \mathbf{N}$ we can find $f_n \in \overline{A_n}^{\circ}$ such that $f_n(x_o) > n$. If $x \in M$ there exists $n_o \in \mathbf{N}$ with $x \in A_{n_o}$, so that $| f_n(x) | \leq 1$ for all $n \geq n_o$. Hence $(f_n |_M)_n$ is $\sigma(M', M)$-bounded and therefore equicontinuous since M is ω-barrelled. Let V be a 0-neighborhood in M such that $| f_n(x) | \leq 1$ for all $x \in V$ and all $n \in \mathbf{N}$. Now $\overline{V}$ is a neighborhood of 0 in E by 1.4.8 (ii) and $| f_n(x) | \leq 1$ for all $x \in \overline{V}$ because each f_n is continuous. Since $\overline{V}$ is absorbing there is an $\epsilon > 0$ with $\epsilon x_o \in \overline{V}$ and therefore $| f_n(x_o) | \leq \epsilon^{-1}$ for all n, a contradiction. It follows that $E = F$.

Let B be a barrel in M and set $A_n = nB$ for $n \in \mathbf{N}$. By what we have just seen, $\overline{B}$ is a barrel and therefore a neighborhood of 0 in E. It follows that $B = \overline{B} \cap M$ is a neighborhood of 0 in M, so that M is barrelled.

Finally, if $(E_n)_n$ is an increasing sequence of subspaces whose union is M, set $A_n = E_n$. Then $E = F = \bigcup_{n\in\mathbf{N}} \overline{E_n}$ and since E is quasi-Baire we infer that $\overline{E_n} = E$ for some $n \in \mathbf{N}$. Hence $\overline{E_n}^M = M$, implying that M is quasi-Baire. ■

6.3.2 Corollary *The completion of a quasi-Baire space is quasi-Baire.* ■

6.3.3 Remark 6.3.1 remains valid if 'quasi-Baire' is replaced by 'Baire-like' as can be seen from 6.2.12 and the proof of 6.3.1 (in its 'if'-part, skip the argument dealing with B and replace $(E_n)_n$ by absolutely convex sets $(A_n)_n$ in M).

6.3.4 Theorem *Every countable-codimensional subspace F of a Baire-like [quasi-Baire] space E is Baire-like [quasi-Baire].*

Proof. It is sufficient to treat the cases of dense resp. closed subspaces: If the codimension of F in E is countable, so are the respective codimensions of F in $\overline{F}$ and of $\overline{F}$ in E.

Let F be dense in E. By 6.3.1 (resp. 6.3.3), F is Baire-like [quasi-Baire] since F is ω-barrelled by 1.7.8.

Now consider the case of F being closed. From 6.2.8 (iii)⇒(i) it follows that the codimension of F in E is finite. [By 1.1.1, F is therefore barrelled]. By finite induction, we may assume that $E = F \oplus sp(x_o)$ for some $x_o \notin F$.

We give the detailed proof for the case of E being Baire-like. The modifications required for the quasi-Baire case will be pointed out afterwards. Suppose that F is not Baire-like and let $(B_n)_n$ be an increasing sequence of rare absolutely convex subsets of F which are closed in F (hence in E) and whose union is F.
Define $A_n := B_n + \{\lambda x_o \mid \mid \lambda \mid \leq n\}$. Each A_n is absolutely convex and closed as the sum of a closed and a compact set. Furthermore, A_n is rare: Assuming A_n to be a 0-neighborhood in E would imply that $B_n = A_n \cap F$ is a 0-neighborhood in F, contradicting its rareness. Since the A_n obviously are increasing and cover E, our hypothesis of F not being Baire-like is absurd.
To obtain the proof for the quasi-Baire case, replace $(B_n)_n$ by an increasing sequence $(F_n)_n$ of closed proper subspaces of F covering F and use $E_n := F_n + sp(x_o)$ instead of A_n to derive the desired contradiction. ■

6.3.5 Theorem *Let $f : E \to F$ be a continuous linear surjection from an lcs E onto an lcs F.*

(i) If E is quasi-Baire and F is barrelled, then F is quasi-Baire.

(ii) If E fails to contain Φ, so does F.

(iii) If E has its weak topology, so does F.

If, additionally, f is nearly open (e.g. when F is barrelled, cf. [50], p.163), then

(iv) If E is Baire-like, so is F.

(v) If E is Baire, so is F.

Proof. (i) Let $(F_n)_n$ be an increasing sequence of closed subspaces covering F. Then $E = \bigcup_{n \in \mathbf{N}} f^{-1}(F_n)$, so that for some n $f^{-1}(F_n) = E$ and therefore $F_n = F$.
(ii) Suppose $\Phi \cong F_o \subseteq F$. Let $\{y_n \mid n \in \mathbf{N}\}$ be a Hamel basis of F_o and choose $\{x_n \mid n \in \mathbf{N}\}$ in E such that $f(x_n) = y_n$ for each $n \in \mathbf{N}$. Then $\{x_n \mid n \in \mathbf{N}\}$ is linearly independent.
Set $E_o := sp(\{x_n \mid n \in \mathbf{N}\})$. Then $f \mid_{E_o}$ is a continuous linear bijection from E_o onto Φ. $(f \mid_{E_o})^{-1} : \Phi \to E_o$ is linear and therefore continuous. Hence $E_o \cong \Phi$.
(iii) An application of 2.3.3: Let U be a neighborhood of zero in F. Then $f^{-1}(U)$ is a neighborhood of 0 in E and therefore contains a finite-codimensional subspace

M. Thus U contains the finite-codimensional subspace $f(M)$.
(iv) Let $(A_n)_n$ be an increasing sequence of closed, absolutely convex subsets covering F. Then since E is covered by $(f^{-1}(A_n))_n$, some $f^{-1}(A_n)$ is a 0-neighborhood in E. Since f is nearly open it follows that $\overline{f(f^{-1}(A_n))} = \overline{A_n} = A_n$ is a 0-neighborhood in F.
(v) If F is covered by the closed sets $\{A_n \mid n \in \mathbf{N}\}$, then $E = \bigcup_{n \in \mathbf{N}} f^{-1}(A_n)$ so that some $f^{-1}(A_n)$ has non-void interior.
Let x be an inner point of $f^{-1}(A_n)$. Then $\overline{f(f^{-1}(A_n) - x)} = A_n - f(x)$ is a neighborhood of 0 in F, showing that the interior of A_n is not empty. ■

6.3.6 Corollary *Let E be an lcs and M a closed subspace of E. Then (i)-(v) of 6.3.5 hold for $F = E/M$. In particular, separated quotients of locally convex Baire spaces, quasi-Baire spaces or Baire-like spaces are Baire, quasi-Baire or Baire-like, respectively.*

Proof. The canonical surjection $\pi : E \to E/M$ is continuous, linear and open and every separated quotient of a barrelled space is barrelled. ■

6.3.7 Proposition *Every finite product $\prod_{i=1}^{n} E_i$ of Baire-like spaces is Baire-like.*

Proof. Using finite induction, it is enough to show that if E_1 and E_2 are Baire-like spaces then $E_1 \times E_2$ is Baire-like. As in the previous chapter we will always identify E_i as a subspace of $E_1 \times E_2$.
Let $(A_n)_n$ be an increasing sequence of closed, absolutely convex sets covering $E_1 \times E_2$. Then $(A_n^{(i)})_n = (A_n \cap E_i)_n$ covers E_i $(i = 1, 2)$. Thus there exists p such that both $A_p^{(1)}$ is a 0-neighborhood in E_1 and $A_p^{(2)}$ is a 0-neighborhood in E_2. Now $\frac{1}{2}A_p^{(1)} \times \frac{1}{2}A_p^{(2)} \subseteq A_p$ is a neighborhood of 0 in $E_1 \times E_2$, i.e. $E_1 \times E_2$ is Baire-like. ■

6.3.8 Theorem *Every arbitrary product $E = \prod_{\iota \in I} E_\iota$ of Baire-like spaces E_ι is Baire-like.*

Proof. To begin with, let $\mid I \mid = \aleph_0$ and let $(A_n)_n$ be an increasing sequence of closed, absolutely convex sets covering E. Without loss of generality we may suppose that I is a dense subset of $(0, 1]$.
Let $D_0 = (0, \frac{1}{2}] \cap I$ and $D_1 = (\frac{1}{2}, 1] \cap I$. Set $D_{00} = (0, \frac{1}{4}] \cap I$, $D_{01} = (\frac{1}{4}, \frac{1}{2}] \cap I$, $D_{10} = (\frac{1}{2}, \frac{3}{4}] \cap I$ and $D_{11} = (\frac{3}{4}, 1] \cap I$. If $D_{r_1 \ldots r_n}$ has been defined for each choice of $r_1, \ldots, r_n \in \{0, 1\}$, set $D_{r_1 \ldots r_n 0} = (a, b] \cap I$, where a is the binary decimal $0.r_1 \ldots r_n 0$

and $b = a + (\frac{1}{2})^{n+1}$. Define $D_{r_1...r_n 1} = (b, c] \cap I$, where $b = 0.r_1...r_n 1$ $(= a + (\frac{1}{2})^{n+1})$ and $c = b + (\frac{1}{2})^{n+1}$.

Then $D_{r_1...r_n 0} \cup D_{r_1...r_n 1} = D_{r_1...r_n}$, $D_{r_1...r_n 0} \cap D_{r_1...r_n 1} = \emptyset$ and the length of $D_{r_1...r_n}$ is $(\frac{1}{2})^n$ for each $n \in \mathbf{N}$.

Suppose no A_n is absorbing in E. With the usual identifications we have

$$E = \prod_{\iota \in D_0} E_\iota \times \prod_{\iota \in D_1} E_\iota.$$

We now use the following fact: If for two vector spaces G_1, G_2 and an absolutely convex set $B \subseteq G_1 \times G_2$, $B \cap G_i$ is absorbing in G_i $(i = 1, 2)$, then B is absorbing in $G_1 \times G_2$. This yields the existence of some $r_1 \in \{0, 1\}$ such that no A_n is absorbing in $\prod_{\iota \in D_{r_1}} E_\iota$. Now

$$\prod_{\iota \in D_{r_1}} E_\iota = \prod_{\iota \in D_{r_1 0}} E_\iota \times \prod_{\iota \in D_{r_1 1}} E_\iota,$$

so that we can inductively define a sequence $(r_n)_n$ in $\{0, 1\}$ such that no A_k is absorbing in $\prod_{\iota \in D_{r_1...r_n}} E_\iota$ for $n = 1, 2....$ Since the diameters of $D_{r_1...r_n}$ form a null sequence, $\bigcap_{n=1}^{\infty} D_{r_1...r_n}$ contains at most one element. This leaves two possibilities:

(i) $\bigcap_{n=1}^{\infty} D_{r_1...r_n} = \emptyset$. Set $n_1 = 1$. Since A_1 is not absorbing in $\prod_{\iota \in D_{r_1}} E_\iota$, there exists $x_1 \in \prod_{\iota \in D_{r_1}} E_\iota \setminus sp(A_1)$. Now $E = \bigcup_{n=1}^{\infty} A_n$, so that $x_1 \in A_{n_2}$ for some $n_2 > n_1$. In this way we can inductively define a sequence $(x_k)_k$ in E and a subsequence $(A_{n_k})_k$ of $(A_n)_n$ such that

$$x_k \in \left(\left(\prod_{\iota \in D_{r_1...r_k}} E_\iota \right) \cap A_{n_{k+1}} \right) \setminus sp(A_{n_k})$$

for each $k \in \mathbf{N}$.

Let q be a continuous seminorm on E. Then there exists a 0-neighborhood $U = \prod_{\iota \in H} U_\iota \times \prod_{\iota \notin H} E_\iota$ (U_ι a 0-neighborhood in E_ι for $\iota \in H$, H finite) such that $q(x) \leq 1$ for all $x \in U$.

Now if $x \in E$ is such that $x_\iota = 0$ for all $\iota \in H$ then clearly $q(x) = 0$. Since $(D_{r_1...r_n})_n$ is a decreasing sequence with empty intersection there exists $n_o \in \mathbf{N}$ with $D_{r_1...r_n} \cap H = \emptyset$ for all $n \geq n_o$. Therefore

$$q\left(\prod_{\iota \in D_{r_1...r_n}} E_\iota \right) = \{0\} \text{ for all } n \geq n_o.$$

In particular, $q(x_k) = 0$ for each $k \geq n_o$. By 6.2.2, $S = sp(\{x_k \mid k \in \mathbf{N}\})$ carries the strongest locally convex topology (E is barrelled as the product of the barrelled spaces E_ι). Hence

$$\widetilde{p} : \sum_{i=1}^{n} \lambda_i x_i \to \sum_{i=1}^{n} \mid \lambda_i \mid$$

defines a continuous seminorm on S. Let p be a continuous seminorm on E with $p \mid_S = \widetilde{p}$ (cf. 2.3.2). Then $p(x_k) = 1$ for each $k \in \mathbf{N}$, a contradiction.

(ii) $\bigcap_{n=1}^{\infty} D_{r_1 \ldots r_n} = \{\iota_o\}$. Since E_{ι_o} is Baire-like, some $A_n \cap E_{\iota_o}$ is absorbing in E_{ι_o}. Therefore no A_n is absorbing in $\prod_{\iota \in D_{r_1 \ldots r_k} \setminus \{\iota_o\}} E_\iota$: Otherwise, by the observation made above, some A_n would be absorbing in

$$\prod_{\iota \in D_{r_1 \ldots r_k}} E_\iota = \left(\prod_{\iota \in D_{r_1 \ldots r_k} \setminus \{\iota_o\}} E_\iota \right) \times E_{\iota_o},$$

a contradiction. As in (i) we can define a sequence $(y_k)_k$ in E and a subsequence $(A_{m_k})_k$ of $(A_n)_n$ such that

$$y_k \in \left(\left(\prod_{\iota \in D_{r_1 \ldots r_k} \setminus \{\iota_o\}} E_\iota \right) \cap A_{m_{k+1}} \right) \setminus sp(A_{m_k})$$

for each $k \in \mathbf{N}$.

Since $\bigcap_{k=1}^{\infty} (D_{r_1 \ldots r_k} \setminus \{\iota_o\}) = \emptyset$, the same argument as in (i) yields a contradiction. Thus we are forced to conclude that some A_n is absorbing in E. Since E is barrelled, A_n is then a 0-neighborhood and E is Baire-like.

Now suppose that $\mid I \mid > \aleph_0$ and that E is covered by an increasing sequence $(A_n)_n$ of closed and absolutely convex sets. Again, assume that no A_n is absorbing.

As in the proof of 6.2.3 we can choose a sequence $(x^{(k)})_k = ((x_\iota^{(k)})_{\iota \in I})_k$ and a subsequence $(A_{n_k})_k$ such that $x^{(k)} \in A_{n_{k+1}} \setminus sp(A_{n_k})$ for each k. Fix $k \in \mathbf{N}$. Since

$$x^{(k)} \notin sp(A_{n_k}) = \bigcup_{j \in \mathbf{N}} j A_{n_k}^{oo},$$

we can choose a sequence $(f_{k_j})_j$ in $A_{n_k}^{o}$ such that $\mid f_{k_j}(x^{(k)}) \mid \geq j$ for each j, i.e. such that $(f_{k_j}(x^{(k)}))_j$ is unbounded.

If $j, k \in \mathbf{N}$, then $x \to \mid f_{k_j}(x) \mid$ is a continuous seminorm on E and as in (i) we conclude that there exists a finite subset $I_{k_j} \subseteq I$ with

$$f_{k_j} \left(\prod_{\iota \in I \setminus I_{k_j}} E_\iota \right) = \{0\}.$$

Set $I_o := \bigcup_{j,k\in\mathbf{N}} I_{k_j}$. I_o is countable and $f_{k_j}(\prod_{\iota\in I\setminus I_o} E_\iota) = \{0\}$ for all k, j. Let

$$\overline{x}^{(k)} = (x_\iota^{(k)})_{\iota\in I_o} \in \prod_{\iota\in I_o} E_\iota =: E_o.$$

Clearly, $f_{k_j}(\overline{x}^{(r)}) = f_{k_j}(x^{(r)})$ for $k, j, r \in \mathbf{N}$. Define

$$B_p := \{f_{k_j} \mid k \geq p, j \in \mathbf{N}\}^\circ \supseteq (A_{n_p}^\circ)^\circ = A_{n_p}.$$

Then $E = \bigcup_{p\in\mathbf{N}} B_p$, $E_o = \bigcup_{p\in\mathbf{N}}(B_p \cap E_o)$ and $(B_p \cap E_o)_p$ is an increasing sequence of absolutely convex sets closed in E_o. For an arbitrary p, $(f_{p_j}(\overline{x}^{(p)}))_j$ is unbounded, so that $B_p \cap E_o$ does not absorb $\overline{x}^{(p)} \in E_o$. Thus every $B_p \cap E_o$ is rare in E_o, contradicting the fact that E_o is Baire-like as a countable product of Baire-like spaces.

As in the former case $|I| = \aleph_0$ we conclude that some A_n is absorbing, hence a barrel and, finally, a 0-neighborhood in E. Therefore E is Baire-like. ■

6.3.9 Proposition *Every finite product $\prod_{i=1}^n E_i$ of quasi-Baire spaces is quasi-Baire.*

Proof. Again, it is enough to show that if E_1 and E_2 are quasi-Baire then $E_1 \times E_2$ is quasi-Baire.

$E_1 \times E_2$ is barrelled. Now let $(F_n)_n$ be an increasing sequence of closed subspaces covering $E_1 \times E_2$. $(F_n \cap E_1)_n$ (resp. $(F_n \cap E_2)_n$) covers E_1 (resp. E_2). Therefore we can find $m, n \in \mathbf{N}$ with $E_1 \subseteq F_m$ and $E_2 \subseteq F_n$. Consequently, $E_1 \times E_2 \subseteq F_{m+n}$ and $E_1 \times E_2$ is quasi-Baire. ■

6.3.10 Theorem *Every arbitrary product $E = \prod_{\iota\in I} E_\iota$ of quasi-Baire spaces E_ι is quasi-Baire.*

Proof. First of all, E is barrelled. Suppose there exists an increasing sequence $(F_n)_n$ of closed, proper subspaces covering E. Denote by $\pi_\iota : E \to E_\iota$ the canonical projection ($\iota \in I$).

$E \setminus F_1$ is non-empty and open, so that there exists a finite subset $H_1 \subseteq I$ and nonempty open sets U_ι in E_ι for $\iota \in H_1$ such that $S_1 := \bigcap_{\iota\in H_1} \pi_\iota^{-1}(U_\iota)$ and F_1 are disjoint. Set $k_1 = 1$.

$E_{H_1} := \prod_{\iota\in H_1} E_\iota$ is quasi-Baire by 6.3.9. As in the proof of 6.3.9 it follows that E_{H_1} is contained in F_{k_2} for some $k_2 > k_1$. Choose a finite subset $\widetilde{H}_2$ of I and open sets $V_\iota \neq \emptyset$ in E_ι ($\iota \in \widetilde{H}_2$) such that $S = \bigcap_{\iota\in\widetilde{H}_2} \pi_\iota^{-1}(V_\iota)$ and F_{k_2} are disjoint.

Set $H_2 := \widetilde{H}_2 \setminus H_1$. Suppose there exists

$$y \in \left(\bigcap_{\iota \in H_2} \pi_\iota^{-1}(V_\iota) \right) \cap F_{k_2}.$$

Choose $z \in E$ with $\pi_\iota(z) = 0$ for $\iota \notin H_1 \cap \widetilde{H}_2$ and $\pi_\iota(z) \in -\pi_\iota(y) + V_\iota$ for $\iota \in H_1 \cap \widetilde{H}_2$. Then $z \in E_{H_1} \subseteq F_{k_2}$ so that $y + z \in F_{k_2}$, contradicting the fact that $y + z \in S$. Hence $S_2 = \bigcap_{\iota \in H_2} \pi_\iota^{-1}(V_\iota)$ and F_{k_2} are disjoint. Set $U_\iota := V_\iota$ for $\iota \in H_2$.
By induction we now obtain disjoint finite subsets $H_1, H_2, \ldots$ of I, non-empty open subsets U_ι in E_ι ($\iota \in \bigcup_{n=1}^\infty H_n$) and a subsequence $(k_n)_n$ of $\mathbf{N}$ such that $S_n = \bigcap_{\iota \in H_n} \pi_\iota^{-1}(U_\iota)$ and F_{k_n} are disjoint for each n. To achieve the latter, choose k_{n+1} in such a way that

$$\prod_{\iota \in H_1 \cup \ldots \cup H_n} E_\iota \subseteq F_{k_{n+1}}.$$

This allows to perform with $H_{n+1} = \widetilde{H}_{n+1} \setminus (H_1 \cup \ldots \cup H_n)$ the analogous steps as above in the special case $n = 1$.
Let x be an element of E with $\pi_\iota(x) \in U_\iota$ for all $\iota \in \bigcup_{n=1}^\infty H_n$. Then for each $n \in \mathbf{N}$, $x \in S_n$ and therefore $x \notin F_{k_n}$. But this contradicts the fact that $E = \bigcup_{n=1}^\infty F_{k_n}$ and we must conclude that E is quasi-Baire. ■

Recall the following notation from chapter 5: If $E = \prod_{\iota \in I} E_\iota$ then E_o denotes the subspace of E consisting of those $x = (x_\iota)_{\iota \in I}$ such that $x_\iota = 0$ for all but countably many $\iota \in I$.

6.3.11 Corollary *Let I be an arbitrary indexing set and for $\iota \in I$ let E_ι be an lcs. If each E_ι is barrelled, quasi-Baire, Baire-like, unordered Baire-like or a Fréchet space, then $E_o \subseteq E = \prod_{\iota \in I} E_\iota$ is, respectively, barrelled, quasi-Baire, Baire-like, unordered Baire-like or a Baire space .*

Proof. By 5.2.17 it is sufficient to prove that E has the respective property. The product of barrelled spaces is barrelled. Any product of Fréchet spaces is Baire by [29], p.43. The other statements follow from 5.2.16, 6.3.8 and 6.3.10. ■

The properties 'Baire-like' and 'quasi-Baire' are, in general, not stable under the formation of inductive limits, as will become clear in chapter 7. There we will also specify large classes of lcs's distinguishing between (db)-, Baire-like- ,quasi-Baire- and barrelled spaces.

6.4 References

6.1.1, 6.1.3 and 6.1.4 were introduced by S.A.SAXON ([45]). 6.1.8 is a perturbation theorem whose original form is due to T.KATO ([25]). It was then successively improved by G.KÖTHE ([30]), M.VALDIVIA ([53]) and P.P.NARAYANASWAMI and S.A.SAXON ([36]).

6.2.1 - 6.2.5 are taken from [45]. 6.2.6 goes back to T. and Y.KOMURA, as was already pointed out in 1.11.

6.2.8 was established by P.P.NARAYANASWAMI and S.A.SAXON ([36]). 6.2.10 - 6.2.13 and 6.3.5 - 6.3.8 are taken from [45].

6.2.14, 6.2.15 and 6.3.11 were obtained by S.A.SAXON and A.R.TODD in [48].

6.3.1, 6.3.2, 6.3.9 and 6.3.10 are again taken from [36].

6.3.4 essentially goes back to [36] and [45].

7 (LF)-Spaces

The various concepts of 'almost Baire' spaces we introduced in the previous chapters provide useful tools for a study of (LF)-spaces. On the one hand, they enable a classification of inductive limits of lcs's. On the other hand, they can be employed to reveal a close relationship between (LF)-spaces and the separable quotient problem in Fréchet and Banach spaces.
This, in turn, leads to the construction of an abundance of metrizable and even normable (LF)-spaces. The final section of this chapter will be concerned with completions of (LF)-spaces.

7.1 Generalities

7.1.1 Definition *We shall say that (E, τ) is the* inductive limit *of the increasing sequence $((E_n, \tau_n))_n$ of locally convex spaces, $(E, \tau) = \underrightarrow{\lim}(E_n, \tau_n)$, if*

(i) $E = \bigcup_{n=1}^{\infty} E_n$.

(ii) $\tau_{n+1} \mid_{E_n} \leq \tau_n$ *for each* $n \in \mathbf{N}$.

(iii) τ *is the strongest locally convex topology on* E *such that* $\tau \mid_{E_n} \leq \tau_n$ *for each* $n \in \mathbf{N}$.

(iv) τ *is Hausdorff.*

$((E_n, \tau_n))_n$ *is called* defining sequence *for* (E, τ).
A defining sequence is called strict *if* $\tau_{n+1} \mid_{E_n} = \tau_n$ *for each* n.
In this case we say that (E, τ) *is the* strict inductive limit *of the sequence* $((E_n, \tau_n))_n$. *(Some authors additionally require* E_n *to be closed in* (E_{n+1}, τ_{n+1}) *for each* n *(cf. e.g. [21], p140).*

7.1.2 Definition *An lcs* (E, τ) *is called* (LF)-space *[*(LB)-space*] if there exists a* strictly *increasing sequence* $((E_n, \tau_n))_n$ *of Fréchet [Banach] spaces such that* $(E, \tau) = \underrightarrow{\lim}(E_n, \tau_n)$.

An (LF)-space [(LB)-space] is called strict *if it has at least one strict defining sequence of Fréchet [Banach] spaces* $((E_n, \tau_n))_n$.

For such a sequence, E_n is necessarily closed in (E_{n+1}, τ_{n+1}) because it is complete. Observe that a strict (LF)-space generally also has many non-strict defining sequences, cf. 7.1.18 and 7.1.21 below, the latter stating that Φ constitutes the only exception to this rule. Compare also 7.1.24.

7.1.3 Remarks

(i) We will see later that, according to the above definition, no Fréchet space is an (LF)-space (cf. 7.2.10).

(ii) Conditions (ii) and (iii) of 7.1.1 mean that the canonical injections $(E_n, \tau_n) \hookrightarrow (E_{n+1}, \tau_{n+1})$ and $(E_n, \tau_n) \hookrightarrow (E, \tau)$ are continuous, i.e. that (E_n, τ_n) is continuously included in (E_{n+1}, τ_{n+1}) and that τ is the strongest locally convex topology such that (E_n, τ_n) is continuously included in (E, τ) for each $n \in \mathbf{N}$.

(iii) When using formulations like 'the (LF)-space $(E, \tau) = \varinjlim(E_n, \tau_n)$' it is always understood that the sequence $((E_n, \tau_n))_n$ just considered is *strictly* increasing. This is no substantial restriction on the choice of $((E_n, \tau_n))_n$ as will become clear in 7.2.10.

7.1.4 First of all, we list some basic properties of inductive limits:

(1) Let $(E, \tau) = \varinjlim(E_n, \tau_n)$
A neighborhood base of the origin in (E, τ) is given by all sets of the form $\Gamma(\bigcup_{n=1}^{\infty} V_n)$, where V_n is a neighborhood of 0 in (E_n, τ_n) for each $n \in \mathbf{N}$. In other words, $U \subseteq E$ is a τ-neighborhood of 0 if and only if $U \cap E_n$ is a τ_n-neighborhood of 0 in E_n for each n (see [23], p.157).
A linear map from E into an lcs F is continuous if and only if $f|_{E_n}: (E_n, \tau_n) \to F$ is continuous for each $n \in \mathbf{N}$ ([23], p.159).

(2) If $(E, \tau) = \varinjlim(E_n, \tau_n)$, then $(E, \tau) = \varinjlim(E_{n_k}, \tau_{n_k})$ for any subsequence $(E_{n_k})_k$ of $(E_n)_n$ by the very definition of the inductive topology.
Thus, either $(E, \tau) = (E_n, \tau_n)$ for some n or (E, τ) possesses a strictly increasing defining sequence.

(3) If all (E_n, τ_n) are barrelled, then $(E, \tau) = \varinjlim(E_n, \tau_n)$ is barrelled by 1.8.2. In particular, every (LF)-space is barrelled.

(4) If $(E,\tau) = \underrightarrow{\lim}(E_n, \tau_n)$ is strict and E_n is closed in (E_{n+1}, τ_{n+1}) for each $n \in \mathbf{N}$, then a subset B of (E,τ) is bounded if and only if B is a bounded subset of some (E_n, τ_n) (see [50], II.6.5).

(5) If (E,τ) is the strict inductive limit of the sequence $((E_n, \tau_n))_n$, this time *without* τ presupposed to be Hausdorff, then $\tau \mid_{E_n} = \tau_n$ for every n and it follows that τ is Hausdorff ([50], II.6.4).

(6) The strict inductive limit of a sequence of complete lcs's is complete ([50], II.6.6). In particular, every strict (LF)-space is complete.

(7) Every closed linear map from an (LF)-space into an (LF)-space is continuous (See [29], p.43).

(8) Every continuous linear surjection from an (LF)-space onto an (LF)-space is open. (See [29], p.43. For a different proof of this fact see 7.3.12 below.)

(9) Two algebraically complementary closed subspaces of an (LF)-space are topologically complementary ([29], p.43).

7.1.5 Example Φ is a strict (LF)-space.

7.1.6 Example Fix $m \in \mathbf{N}$ and let $(K_n)_n$ be a sequence of compact subsets of $\mathbf{R}^m$ covering $\mathbf{R}^m$ such that $K_n \subseteq int(K_{n+1})$ for $n = 1, 2, \ldots$. It follows that each compact subset of $\mathbf{R}^m$ is contained in some K_n: Observe that

$$\mathbf{R}^m = \bigcup_{n=1}^{\infty} K_n \subseteq \bigcup_{n=1}^{\infty} int(K_{n+1}).$$

Hence every compact $K \subseteq \mathbf{R}^m$ is covered by finitely many $int(K_n)$ and since $(K_n)_n$ is increasing, our claim is proved.
Denote by $\mathcal{D}(K_n)$ the vector space of all infinitely differentiable complex valued functions on $\mathbf{R}^m$ whose support is contained in K_n. With the generating family

$$p_k(f) = \sup_{x \in \mathbf{R}^m} \mid D^k f(x) \mid \quad (k \in \mathbf{N}),$$

$\mathcal{D}(K_n)$ is a Fréchet space.
Let $\mathcal{D}$ be the vector space of all infinitely differentiable functions on $\mathbf{R}^m$ with compact support. Then $\mathcal{D} = \bigcup_{n=1}^{\infty} \mathcal{D}(K_n)$ and we equip $\mathcal{D}$ with the strongest locally

convex topology τ such that $\mathcal{D}(K_n)$ is continuously included in $\mathcal{D}$ for each n.
Since for compact sets $K' \subseteq K''$, $\mathcal{D}(K')$ is continuously included in $\mathcal{D}(K'')$, τ is also the finest locally convex topology on $\mathcal{D}$ for which *all* inclusions $\mathcal{D}(K) \hookrightarrow \mathcal{D}$ (K a compact subset of $\mathbf{R}^m$) are continuous.
$(\mathcal{D}, \tau)$ is separated since τ is finer than the (Hausdorff) topology of uniform convergence on all compact subsets of $\mathbf{R}^m$. Clearly $\mathcal{D}(K_{n+1})$ induces on $\mathcal{D}(K_n)$ its own topology for each n. Hence $(\mathcal{D}, \tau)$ is a strict (LF)-space. The elements of $(\mathcal{D}, \tau)'$ are called complex distributions on $\mathbf{R}^m$ (cf. [23]).

7.1.7 Example Let $C_{oo}(\mathbf{R}^m)$ be the complex vector space of all continuous functions on $\mathbf{R}^m$ with compact support. With $(K_n)_n$ as in 7.1.6 we have

$$C_{oo}(\mathbf{R}^m) = \bigcup_{n=1}^{\infty} C_{oo}(K_n),$$

where $C_{oo}(K_n)$ is the space of all continuous functions on $\mathbf{R}^m$ whose support is contained in K_n.
Each $C_{oo}(K_n)$ is a Banach space with

$$\|f\| := \sup_{x \in K_n} | f(x) |$$

and we equip $C_{oo}(\mathbf{R}^m)$ with the finest locally convex topology τ such that all inclusions $C_{oo}(K_n) \hookrightarrow C_{oo}(\mathbf{R}^m)$ are continuous.
As in 7.1.6, τ is the finest locally convex topology on $C_{oo}(\mathbf{R}^m)$ for which every $C_{oo}(K)$ ($K \subseteq \mathbf{R}^m$ compact) is continuously included in $C_{oo}(\mathbf{R}^m)$. Again, τ is Hausdorff because it is finer than the topology of uniform convergence on all compact subsets of $\mathbf{R}^m$.
Finally, since each $C_{oo}(K_n)$ can be considered as a topological subspace of $C_{oo}(K_{n+1})$, $C_{oo}(\mathbf{R}^m)$ is a strict (LB)-space.
The dual of $C_{oo}(\mathbf{R}^m)$ is the space of all complex measures on $\mathbf{R}^m$ (cf. [7]).

We will repeatedly need the following corollary of a well known open mapping theorem by PTÀK:

7.1.8 Proposition *Every continuous linear surjection from a Fréchet space onto a* *barrelled space is open.*

Proof. See [23], p.299, Prop.2. ■

7.1.9 Theorem *Let $(E,\tau) = \varinjlim(E_n, \tau_n)$, where $((E_n, \tau_n))_n$ is a sequence of Fréchet spaces. The following statements are equivalent:*

(i) $\tau_{n+1} |_{E_n} = \tau_n$ for each n, i.e. (E,τ) is the strict inductive limit of $((E_n, \tau_n))_n$.

(ii) $\tau |_{E_n} = \tau_n$ for each n.

(iii) E_n is τ_{n+1}-closed in E_{n+1} for each n.

(iv) E_n is τ-closed in E for each n.

(v) Every $(E_n, \tau |_{E_n})$ is complete.

(vi) Every $(E_n, \tau |_{E_n})$ is sequentially complete.

Proof. (i)⇒(ii) See [50], II.6.4.
(ii)⇒(iv) $(E_n, \tau |_{E_n}) = (E_n, \tau_n)$ is complete and hence closed in (E,τ).
(iv)⇒(iii) E_n is closed in $(E_{n+1}, \tau |_{E_{n+1}})$ and therefore in (E_{n+1}, τ_{n+1}) since $\tau |_{E_{n+1}} \leq \tau_{n+1}$.
(iii)⇒(i) $id : (E_n, \tau_n) \to (E_n, \tau_{n+1} |_{E_n})$ is a continuous linear bijection. By our hypothesis, both (E_n, τ_n) and $(E_n, \tau_{n+1} |_{E_n})$ are Fréchet. Thus 7.1.8 yields $\tau_n = \tau_{n+1} |_{E_n}$.
(ii)⇒(v)⇒(vi) Clear.
(vi)⇒(iii) Let $x = \tau_{n+1}\text{-}\lim x_n$, where $(x_n)_n$ is a sequence in E_n. Since $\tau |_{E_{n+1}} \leq \tau_{n+1}$ it follows that $x = \tau |_{E_{n+1}}\text{-}\lim x_n$.
In particular, $(x_n)_n$ is a $\tau |_{E_{n+1}}$-Cauchy sequence. Consequently, $(x_n)_n$ is a $\tau |_{E_n}$-Cauchy sequence and therefore $x \in E_n$. Since (E_{n+1}, τ_{n+1}) is metrizable, the proof is complete. ■

7.1.10 Proposition *Let E be a vector space, $((E_n, \tau_n))_n$ a sequence of Fréchet spaces and $f_n : E \to E_n$ a linear mapping for each $n \in \mathbf{N}$. Furthermore, let (X,τ) be some Hausdorff topological space and $g_n : (E_n, \tau_n) \to (X,\tau)$ a continuous map for each n, i.e.*

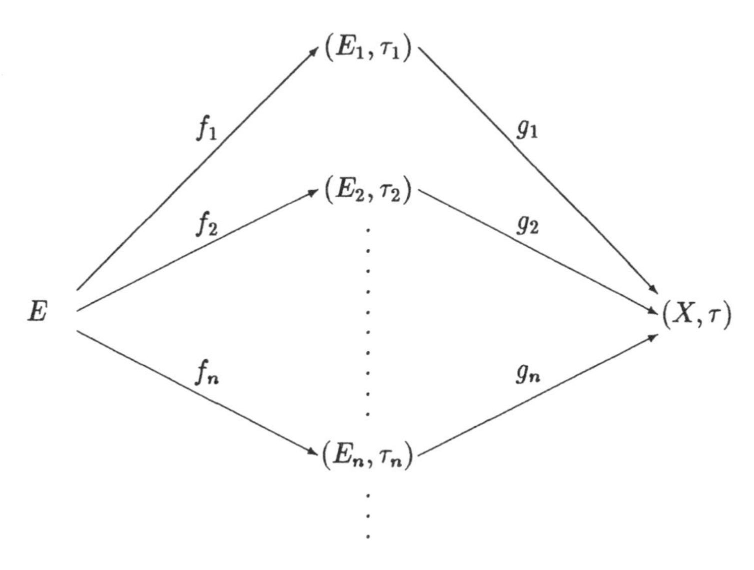

Assume that $g_n \circ f_n = g_m \circ f_m$ for all $n, m \in \mathbf{N}$ and, conversely, that each sequence $(y_n)_n$ such that $g_n(y_n)$ is one and the same element $z \in X$ for all n is of the form $(y_n)_n = (f_n(x))_n$ for some $x \in E$.
Then (E, σ) is a Fréchet space, where σ denotes the projective topology with respect to the mappings f_n, $n \in \mathbf{N}$.

Proof. A neighborhood base of zero for σ is given by the family of all sets U ($\subseteq E$) which are finite intersections of sets of the form $f_n^{-1}(U_n)$, where U_n is a member of a 0-neighborhood base in some (E_n, τ_n) ([50], p. 51). Hence (E, σ) is metrizable.
If $(x_k)_k$ is a Cauchy sequence in (E, σ), then for each n, $(f_n(x_k))_k$ is a Cauchy sequence in (E_n, τ_n), hence convergent to some $y_n \in E_n$.
Since $g_n(y_n) = \lim_{k\to\infty} g_n(f_n(x_k))$ is independent of n, there exists some $x \in E$ such that $y_n = f_n(x)$ for $n \in \mathbf{N}$.
From $f_n(x) = y_n = \lim_{k\to\infty} f_n(x_k)$ it follows that $(x_k - x)_k$ finally is in each set U as described above, i.e. $x = \lim_{k\to\infty} x_k$ with respect to σ. ■

7.1.11 Remarks

(i) Under the assumption of 7.1.10, $h := g_1 \circ f_1$ ($= g_n \circ f_n$ for each $n \in \mathbf{N}$) is continuous from (E, σ) into (X, τ).

(ii) 7.1.10 is also valid for finite families $((E_n, \tau_n))_n$, $(f_n)_n$, $(g_n)_n$: just repeat the last member of each of these finite sequences infinitely many times.

7.1.12 In the sequel, we will mainly consider the following two particular cases of 7.1.10:

(i)

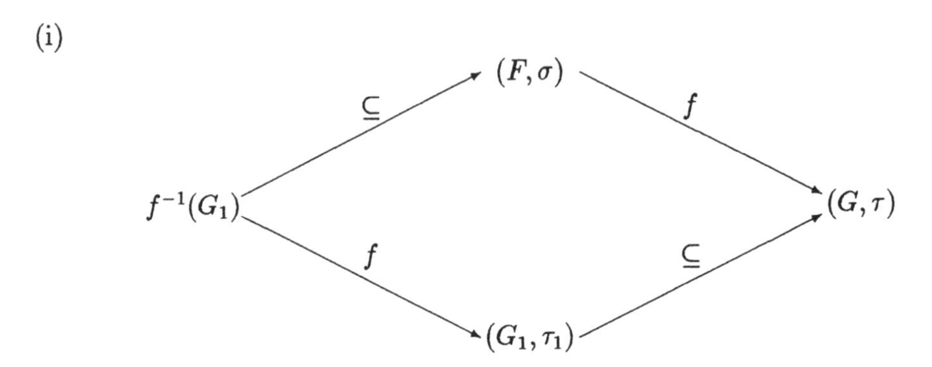

Here, f is a continuous linear map from a Fréchet space (F,σ) into a Fréchet space (G,τ) and (G_1,τ_1) is a Fréchet space continuously included in (G,τ).

(ii)

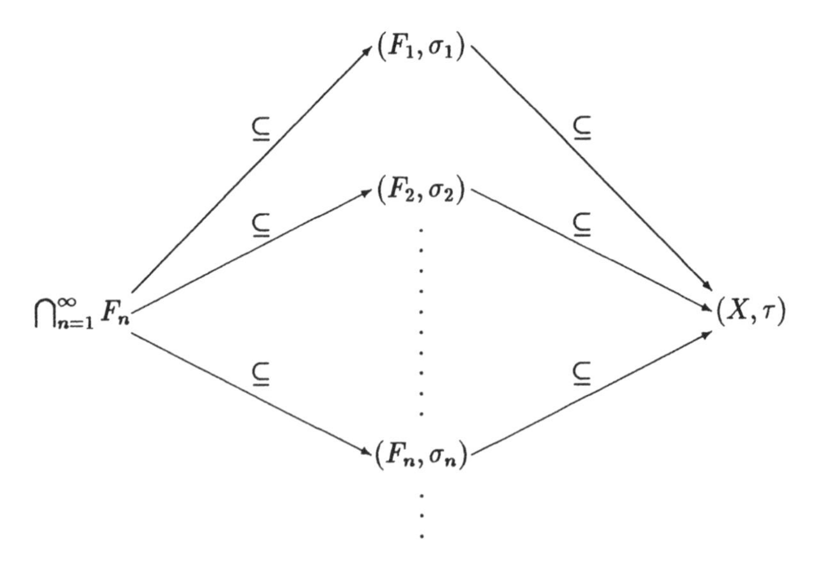

Here, each (F_n,σ_n) is a Fréchet space continuously included in some Hausdorff topological space (X,τ).

7.1.13 Definition *Let $((E_n^{(1)},\tau_n^{(1)}))_n$ and $((E_n^{(2)},\tau_n^{(2)}))_n$ be two sequences of subspaces of the same vector space E such that $(E,\tau^{(i)}) = \varinjlim(E_n^{(i)},\tau_n^{(i)})$ for $i = 1,2$. The sequences $((E_n^{(1)},\tau_n^{(1)}))_n$ and $((E_n^{(2)},\tau_n^{(2)}))_n$ are called* equivalent *if each member* *of one sequence is continuously included in some member of the other sequence and vice versa.*

One of the most important tools in this chapter is the following proposition which is due to GROTHENDIECK:

7.1.14 Proposition *Let (E,σ) be a Fréchet space and $((F_n,\tau_n))_n$ a sequence of Fréchet spaces continuously included in some lcs (F,τ) such that $F = \bigcup_{n=1}^{\infty} F_n$. If $u : (E,\sigma) \to (F,\tau)$ is linear and continuous then for some $n \in \mathbf{N}$, $u(E) \subseteq F_n$ and u is continuous with respect to σ and τ_n.*

Proof. The Fréchet (hence unordered Baire-like) space (E,σ) is the union of the sequence of subspaces $E_n := u^{-1}(F_n)$. By 5.1.4 there exists some $n \in \mathbf{N}$ such that $(E_n, \sigma\mid_{E_n})$ is barrelled and dense in (E,σ).
By 7.1.12 (i), (E_n,ξ) is a Fréchet space, where ξ denotes the projective topology with respect to the inclusion $E_n \hookrightarrow (E,\sigma)$ and the map $u\mid_{E_n}: E_n \to (F_n,\tau_n)$. From 7.1.8 we infer that $id : (E_n,\xi) \to (E_n,\sigma\mid_{E_n})$ is a topological isomorphism.
Therefore, $E_n = E$ ($(E_n,\sigma\mid_{E_n})$ being complete, hence closed in (E,σ)) and $\xi = \sigma$ which yields the continuity of u as a map from $(E,\sigma) = (E_n,\xi)$ into (F_n,τ_n). ∎

In particular, 7.1.14 applies to a Fréchet space continuously embedded in an (LF)-space. Compare 7.1.14 to 6.1.4 and 6.1.5.

7.1.15 Theorem *Let $(E,\tau^{(i)}) = \underrightarrow{\lim}(E_n^{(i)},\tau_n^{(i)})$ for $i = 1,2$, where each $(E_n^{(i)},\tau_n^{(i)})$ is a Fréchet space. The following statements are equivalent:*

(i) $((E_n^{(1)},\tau_n^{(1)}))_n$ is equivalent to $((E_n^{(2)},\tau_n^{(2)}))_n$.

(ii) $\tau^{(1)} = \tau^{(2)}$.

(iii) The infimum of $\tau^{(1)}$ and $\tau^{(2)}$ is Hausdorff.

Proof. (i)⇒(ii) For an arbitrary $n \in \mathbf{N}$, choose $k \in \mathbf{N}$ such that $(E_n^{(1)},\tau_n^{(1)})$ is continuously included in $(E_k^{(2)},\tau_k^{(2)})$.
Since $(E_k^{(2)},\tau_k^{(2)}) \hookrightarrow (E,\tau^{(2)})$ is continuous, it follows that $(E_n^{(1)},\tau_n^{(1)})$ is continuously included in $(E,\tau^{(2)})$. Thus $\tau^{(2)} \leq \tau^{(1)}$ by the definition of $\tau^{(1)}$ as inductive topology. Since the above argument is symmetric in $i \in \{1,2\}$, the assertion is proved.
(ii)⇒(iii) Clear.
(iii)⇒(i) Set $\tau := \inf(\tau^{(1)},\tau^{(2)})$, i.e. let τ be the finest topology coarser than $\tau^{(1)}$ and $\tau^{(2)}$. Then $(E_n^{(i)},\tau_n^{(i)})$ is continuously included in (E,τ) for $i = 1,2$ and $n \in \mathbf{N}$.

Fix $n \in \mathbf{N}$.
By 7.1.14 there exists $p \in \mathbf{N}$ such that $(E_n^{(1)}, \tau_n^{(1)})$ is continuously included in $(E_p^{(2)}, \tau_p^{(2)})$. Due to the symmetry of the proof in $i \in \{1,2\}$ it follows that the sequences $((E_n^{(1)}, \tau_n^{(1)}))_n$ and $((E_n^{(2)}, \tau_n^{(2)}))_n$ are equivalent. ■

7.1.16 Corollary *For an lcs (E, τ) there exists at most one topology $\sigma \geq \tau$ on E such that (E, σ) is an (LF)-space.* ■

7.1.17 Corollary *Let the (LF)-space (E, τ) be the strict limit of the Fréchet spaces $((E_n, \tau_n))_n$ and also an (LB)-space: $(E, \tau) = \underrightarrow{\lim}(E_n, \tau_n) = \underrightarrow{\lim}(B_n, \sigma_n)$, where $((B_n, \sigma_n))_n$ denotes a defining sequence of Banach spaces. Then each (E_n, τ_n) is a Banach space.*

Proof. By 7.1.15, $((B_n, \sigma_n))_n$ and $((E_n, \tau_n))_n$ are equivalent. Thus for each $k \in \mathbf{N}$ there exist l and m with $E_k \subseteq B_l \subseteq E_m$, $\tau_m |_{B_l} \leq \sigma_l$ and $\sigma_l |_{E_k} \leq \tau_k$, implying $\tau_m |_{E_k} \leq \sigma_l |_{E_k} \leq \tau_k$.
Since (E, τ) is the strict inductive limit of $((E_n, \tau_n))_n$, $\tau_m |_{E_k} = \tau_k$. Hence $\tau_k = \sigma_l |_{E_k}$ is normable. ■

7.1.18 Example The non-normable Fréchet space (s) of all rapidly decreasing sequences is continuously included in l^1. For $n \in \mathbf{N}$, set

$$E_n := \underbrace{l^1 \times ... \times l^1}_{n} \times \{0\} \times \{0\} \times ... \text{ and } F_n := \underbrace{l^1 \times ... \times l^1}_{n} \times (s) \times \{0\} \times \{0\} \times ...$$

and equip E_n and F_n with the product topologies σ_n and τ_n, respectively.
Then $E_n \hookrightarrow F_n \hookrightarrow E_{n+1}$ is continuous for every n and therefore $((E_n, \sigma_n))_n$ and $((F_n, \tau_n))_n$ are equivalent sequences in $E = \bigcup_{n=1}^{\infty} E_n = \bigcup_{n=1}^{\infty} F_n$. Since $((E_n, \sigma_n))_n$ is strict, $(E, \tau) = \underrightarrow{\lim}(E_n, \sigma_n)$ is a strict (LB)-space.
By 7.1.15 we also have $(E, \tau) = \underrightarrow{\lim}(F_n, \tau_n)$, but now $((F_n, \tau_n))_n$ is a non-strict defining sequence of non-Banach spaces.
Replacing l^1 by l^2 and (s) by l^1 yields a strict (LB)-space having also a non-strict defining sequence of Banach spaces.

In view of 7.1.18 the question arises if there is an internal characterization of those (LF)-spaces for which every defining sequence is strict. 7.1.21 and 7.1.24 will completely clarify the situation.

7.1.19 Lemma *Let (E, τ) be an infinite-dimensional Fréchet space. Then there exists a subspace F of E and a topology τ_F on F such that (F, τ_F) is a Fréchet space* *and* $\tau\ |_F < \tau_F$.

Proof. Choose a countable 0-neighborhood base $(V_n)_n$ in (E,τ) with $V_{n+1} \subseteq V_n$ for $n \in \mathbf{N}$. Since $dim(E) = \infty$, we can inductively choose $x_n \in V_n \setminus sp(\{x_1, ..., x_{n-1}\})$ for each $n \in \mathbf{N}$.
The sequence $(x_n)_n$ converges to 0, so that $\{x_n \mid n \in \mathbf{N}\} \cup \{0\}$ is compact. Hence $B := \overline{\Gamma}\{x_n \mid n \in \mathbf{N}\}$ is absolutely convex and compact ([50], II.4.3). Denote by E_B the normed space $sp(B)$ with unit ball B.
E_B is a Banach space ([23], p.207) which is continuously included in (E,τ) ([23], p.208), so that $\tau_{E_B} \geq \tau\ |_{E_B}$. Moreover, E_B is infinite-dimensional because $\{x_n \mid n \in \mathbf{N}\} \subseteq E_B$. Therefore $\tau_{E_B} \neq \tau\ |_{E_B}$. (Otherwise B would be a compact neighborhood of zero in E_B.) ■

7.1.20 Remarks

(i) As the proof shows, F can be chosen to be a Banach space.

(ii) The codimension of F in E is necessarily infinite:
Suppose $dim(E/F) < \infty$. Then some finite-dimensional subspace M of E is an algebraic complement of F in E. Now

$$(F,\tau_F) \oplus M \xrightarrow{id} (F,\tau_E\ |_F) \oplus M \xrightarrow{+} (E,\tau_E)$$

is continuous and therefore $E \cong (F,\tau_F) \oplus M$ by 7.1.8. Hence F is closed in E, so that $(F,\tau\ |_F)$ is Fréchet. Again by 7.1.8 we conclude that $\tau_F = \tau\ |_F$, a contradiction.

(iii) Remark (ii) (and several other applications to come) suggest to explicitly state the following observation:
If $E_1, ..., E_n$ are lcs's, then

$$\prod_{i=1}^n E_i = \bigoplus_{i=1}^n E_i,$$

i.e. the product topology on $\prod_{i=1}^n E_i$ is simultaneously the coarsest topology for which the projections $\pi_j : \prod_{i=1}^n E_i \to E_j$ $(1 \leq j \leq n)$ are continuous and the finest locally convex topology for which the embeddings $j_k : E_k \hookrightarrow \bigoplus_{i=1}^n E_i$ $(1 \leq k \leq n)$ are continuous.

Proof. For $1 \leq i \leq n$, let U_i be an absolutely convex neighborhood of 0 in

E_i. Then

$$\frac{1}{n}\prod_{i=1}^{n} U_i \subseteq \Gamma_{i=1}^{n} j_i(U_i) \subseteq \prod_{i=1}^{n} U_i,$$

which proves our assertion. ■

Especially when dealing with the question of interchanging inductive limits it is useful to view finite products of lcs's as locally convex direct sums.

7.1.21 Theorem *Up to isomorphism, Φ is the only (LF)-space for which every defining sequence is strict.*

Proof. Suppose $(\Phi, \tau) = \varinjlim(E_n, \tau_n)$. Then $\tau|_{E_n} \leq \tau_n$ for each n. On the other hand, $\tau|_{E_n}$ is the finest locally convex topology on E_n (cf. the last item of 2.1), so that $\tau|_{E_n} \geq \tau_n$. By 7.1.9, we are done. (As a matter of fact, as the remaining part of the proof will show, each E_n is finite-dimensional and therefore carries a unique locally convex topology.)
Conversely, let (E, τ) be an (LF)-space for which every defining sequence is strict and let $(E, \tau) = \varinjlim(E_n, \tau_n)$. We claim that each E_n is finite-dimensional.
Otherwise, without loss of generality we may assume that $dim(E_1) = \infty$ (cf. 7.1.4 (2)). By 7.1.19 there exists a Fréchet space $(E_o, \tau_o) \subseteq E_1$ such that $\tau_1|_{E_o} < \tau_o$. But then $(E, \tau) = \varinjlim_{n\geq 0}(E_n, \tau_n)$ and $((E_n, \tau_n))_{n\geq 0}$ is not strict, a contradiction. Consequently, $dim(E) = \aleph_0$.
Moreover, by its definition, τ is the finest locally convex topology on E such that $(E_n, \tau_n) \hookrightarrow (E, \tau)$ is continuous for each n. But since the E_n are finite-dimensional, these inclusions are continuous for *every* locally convex topology on E. Hence τ is the strongest locally convex topology on E, i.e. $(E, \tau) \cong \Phi$. (One could also argue that E, being an (LF)-space is barrelled and use 2.2.2.) ■

7.1.22 Lemma *Let (E, τ_E) and (F, τ_F) be Fréchet spaces such that E is continuously included in F and $dim(F/E) = \infty$. Then there exists a Fréchet space (G, τ_G) such that $E \subseteq G \subseteq F$, $\tau_G|_E \leq \tau_E$ and $\tau_F|_G < \tau_G$.*

Proof. If $dim(F/\overline{E}^F) < \infty$, then $E \neq \overline{E}^F$ and therefore $\tau_E > \tau_F|_E$ (otherwise $(E, \tau_E) = (E, \tau_F|_E)$ would be complete and hence closed in F). Thus we can set $G = E$ in this case.
Now suppose $dim(F/\overline{E}^F) = \infty$. By 7.1.19 there exists a Fréchet space $(G_o, \tau_{G_o}) \subseteq F/\overline{E}^F$ with $\tau_{F/\overline{E}}|_{G_o} < \tau_{G_o}$. Furthermore, the codimension of G_o in $F/\overline{E}^F$ is infinite

(see 7.1.20 (ii)).

Denote by π the canonical surjection $F \to F/\overline{E}^F$ and equip $G := \pi^{-1}(G_o)$ with τ_G, the projective topology with respect to the embedding $G \hookrightarrow F$ and the mapping $\pi_o = \pi|_G: G \to G_o$. Now 7.1.12 (i), applied to

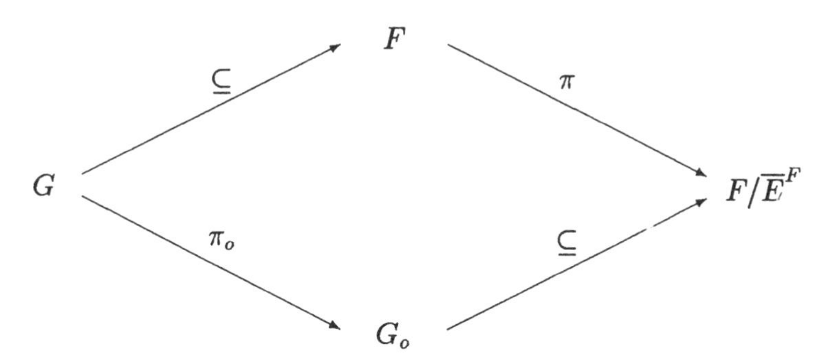

yields that (G, τ_G) is a Fréchet space.

Let V be a neighborhood of 0 in G_o which is not a $\tau_{F/\overline{E}}|_{G_o}$-neighborhood of 0. Then $\pi^{-1}(V)$ is a 0-neighborhood in G and we claim that it is not a $\tau_F|_G$-neighborhood of 0.

Indeed, suppose there exists a 0-neighborhood W in F with $W \cap G \subseteq \pi^{-1}(V)$. Then $\pi(W)$ is a 0-neighborhood in $F/\overline{E}$ and

$$\pi(W) \cap G_o = \pi(W \cap \pi^{-1}(G_o)) = \pi(W \cap G) \subseteq \pi(\pi^{-1}(V)) = V,$$

contradicting the choice of V. It follows that $\tau_G > \tau_F|_G$ and the proof is complete.■

7.1.23 Remark In the preceding proof, $\overline{E} = \pi^{-1}(\{0\}) \subseteq G$. Moreover, the embedding of $(\overline{E}, \tau_F|_{\overline{E}})$ into (G, τ_G) is continuous by the universal property of projective topologies because $\overline{E} \hookrightarrow F$ and $\pi_o|_{\overline{E}} = 0$ are continuous.

7.1.24 Theorem *For an (LF)-space (E, τ), the following statements are equivalent:*

(i) If $((E_n, \tau_n))_n$ is any defining sequence for (E, τ), there exists some $n_o \in \mathbf{N}$ such that $\tau_{n+1}|_{E_n} = \tau_n$ for all $n \geq n_o$ (i.e. every defining sequence is almost strict).

(ii) If $((E_n, \tau_n))_n$ is any defining sequence for (E, τ), there exists some $n \in \mathbf{N}$ such that $\tau_{n+1}|_{E_n} = \tau_n$ (i.e. $((E_n, \tau_n))_n$ is 'strict at one step $E_n \subseteq E_{n+1}$', at least).

(iii) If $((E_n, \tau_n))_n$ is any defining sequence for (E, τ), there exists some $n_o \in \mathbf{N}$ such that E_{n+1}/E_n is finite-dimensional for all $n \geq n_o$.

(iv) *There exists a defining sequence* $((E_n, \tau_n))_n$ *for* (E, τ) *such that* E_{n+1}/E_n *is finite-dimensional for almost all* $n \in \mathbf{N}$.

(v) (E, τ) *is isomorphic to* $F \oplus \Phi$ *for some Fréchet space* F.

Proof. (i)⇒(ii) Clear.
(ii)⇒(iii) Suppose there exists a defining sequence $((E_n, \tau_n))_n$ for (E, τ) such that $dim(E_{n+1}/E_n) = \infty$ for infinitely many n. By 7.1.4 (2) we may assume that for each $n \in \mathbf{N}$, $dim(E_{n+1}/E_n) = \infty$.
Consider those pairs (E_n, E_{n+1}) such that $\tau_{n+1} |_{E_n} = \tau_n$. For such n, 7.1.22 yields the existence of a Fréchet space (F_n, σ_n) with $E_n \subsetneq F_n \subsetneq E_{n+1}$, $\sigma_n |_{E_n} \leq \tau_n$ and $\tau_{n+1} |_{F_n} < \sigma_n$. Moreover, both inclusion are necessarily proper: If $E_n = F_n$, then $\tau_n = \sigma_n$ by 7.1.8 and therefore $\sigma_n = \tau_n = \tau_{n+1} |_{E_n}$, a contradiction. If $E_{n+1} = F_n$, then $\tau_{n+1} = \sigma_n$ by 7.1.8, again providing a contradiction.
For every strict step (E_n, E_{n+1}), place F_n between E_n and E_{n+1}. This way we obtain a strictly increasing sequence of subspaces whose union is E, inducing the same limit topology on E as $((E_n, \tau_n))_n$ (again by 7.1.4 (2)).
Now delete from this new defining sequence all 'left members of strict steps' $E_n \subsetneq E_{n+1}$, i.e. all those E_n such that $\tau_{n+1} |_{E_n} = \tau_n$. The resulting sequence $((G_n, \rho_n))_n$, again defining the same inductive limit topology on E as $((E_n, \tau_n))_n$, differs from $((E_n, \tau_n))_n$ insofar as each (E_n, τ_n) being a left member of a strict step in $((E_n, \tau_n))_n$ has been replaced by the corresponding (F_n, σ_n).
We claim that $\rho_{n+1} |_{G_n} < \rho_n$ for all $n \in \mathbf{N}$ (which provides a contradiction to (v)). For (G_n, ρ_n), (G_{n+1}, ρ_{n+1}), we have to consider four cases:

(a) (E_n, τ_n), (E_{n+1}, τ_{n+1}): the occurence of E_n in $(G_k)_k$ signifies that $\tau_{n+1} |_{E_n} < \tau_n$.

(b) (E_n, τ_n), (F_{n+1}, σ_{n+1}): Similarly to (a) we have $\sigma_{n+1} |_{E_n} \leq \tau_{n+1} |_{E_n} < \tau_n$.

(c) (F_n, σ_n), (E_{n+1}, τ_{n+1}): By construction of (F_n, σ_n), $\tau_{n+1} |_{F_n} < \sigma_n$.

(d) (F_n, σ_n), (F_{n+1}, σ_{n+1}): Again, $\sigma_{n+1} |_{F_n} \leq \tau_{n+1} |_{F_n} < \sigma_n$.

(iii)⇒(iv) Clear.
(iv)⇒(v) Let $(E, \tau) = \underrightarrow{\lim}(E_n, \tau_n)$, where $dim(E_{n+1}/E_n) < \infty$ for each n. E_n is necessarily closed in E_{n+1} $(n = 1, 2, ...)$, i.e. the limit is strict (cf. the proof of

7.1.20 (ii)). For each $n \in \mathbf{N}$ we have $E_{n+1} = E_1 \oplus N_n$, where N_n is a finite-dimensional algebraic (and hence topological) complement of E_1 in E_{n+1}. Thus $(E_{n+1}, \tau_{n+1}) \cong (E_1, \tau_1) \oplus N_1$ for every $n \in \mathbf{N}$. Consequently,

$$(E, \tau) \cong \underrightarrow{\lim}(E_1 \oplus N_n) = E_1 \oplus \underrightarrow{\lim} N_n = E_1 \oplus \Phi$$

(cf. 7.3.5).
(v)⇒(i) Let $(E, \tau) = \underrightarrow{\lim}(E_n, \tau_n) = F \oplus \Phi$ and let $\Phi = \underrightarrow{\lim}_{n\geq 0} M_n$, where $dim(M_n) = n$ for $n \in \mathbf{N}$ and $M_o = \{0\}$.
Then $(E, \tau) = \underrightarrow{\lim}_{n\geq 0}(F \oplus M_n)$ (again by 7.3.5) and 7.1.15 yields the existence of some $n_o \in \mathbf{N}$ such that $F \oplus M_o = F \oplus \{0\}$ is continuously included in E_{n_o}. If n is any natural number $\geq n_o$, there exists some $k_n \in \mathbf{N}$ such that E_n is continuously included in $F \oplus M_{k_n}$, i.e. $\tau_n \geq (\tau_F \oplus \tau_{k_n}) \mid_{E_n}$ ($\tau_F \oplus \tau_{k_n}$ denoting the product topology, cf. 7.1.20 (iii)). Thus

$$\tau_F \oplus \tau_{\{0\}} \geq \tau_n \mid_{F\oplus\{0\}} \geq (\tau_F \oplus \tau_{k_n}) \mid_{F\oplus\{0\}} = \tau_F \oplus \tau_{\{0\}},$$

so that $\tau_n \mid_{F\oplus\{0\}} = \tau_F$ for every $n \geq n_o$. It follows that, for each $n \geq n_o$, F is a (complete and therefore closed) subspace of E_n.
Let G_n be an algebraic complement of F in E_n ($n = 1, 2, ...$). Since $E_n \subseteq F \oplus M_{k_n}$, $dim(G_n) \leq k_n < \infty$. Hence E_n is topologically isomorphic to $F \oplus G_n$ and $G_n \subseteq G_{n+1}$ for $n \geq n_o$. Therefore (E_n, τ_n) is a subspace of (E_{n+1}, τ_{n+1}), i.e. $\tau_{n+1} \mid_{E_n} = \tau_n$ for each $n \geq n_o$. ■

7.2 Classification

The following is a most useful definition due to P.P.NARAYANASWAMI and S.A. SAXON.

7.2.1 Definition *An (LF)-space (E, τ) is of type i or an $(LF)_i$-space ($i = 1, 2, 3$), if it satisfies the following condition (i):*

(1) E has a defining sequence $(E_n)_n$ such that no E_n is dense in E.

(2) E is not metrizable and has a defining sequence $(E_n)_n$ such that some E_n is dense in E.

(3) E is metrizable.

7.2.2 Remarks

(i) By 7.1.4 (2), condition (2) is equivalent to

(2′) *E is not metrizable and has a defining sequence $(E_n)_n$ such that each E_n is dense in E.*

(ii) To see that 7.2.1 actually provides a partition of all (LF)-spaces we have to show that the three classes are pairwise disjoint:
Clearly, $(LF)_2 \cap (LF)_3 = \emptyset$. By 7.1.15, $(LF)_1 \cap (LF)_2 = \emptyset$. Finally, suppose there exists $E \in (LF)_1 \cap (LF)_3$ and let $(E_n)_n$ be a defining sequence for E such that no E_n is dense in E. Since E is metrizable and barrelled, it is Baire-like by 6.2.6. But then some E_n is not rare in E and therefore dense, a contradiction.

(iii) From the reasoning of (ii) and by 7.1.15, we see that the classification of 7.2.1 can also be given the following form:
An (LF)-space (E, τ) is of

-type 1 , if in every defining sequence $((E_n, \tau_n))_n$, each E_n is not dense in (E, τ).

-type 2 , if in every defining sequence $((E_n, \tau_n))_n$, almost all E_n are dense in (E, τ) and (E, τ) is not metrizable.

-type 3 , if in every defining sequence $((E_n, \tau_n))_n$, almost all E_n are dense in (E, τ) and (E, τ) is metrizable.

(iv) In [36], p.629, NARAYANASWAMI and SAXON concede the possibility to integrate Fréchet spaces in the above list as $(LF)_4$-spaces (rather than as $(LF)_o$-spaces because of 7.3.11). However, to do this, one would have to relax the requirement that every defining sequence has to be strictly increasing which would spoil many properties peculiar to (LF)-spaces. Therefore we will not adopt this terminology.

(v) Every strict (LF)-space is of type 1 by 7.1.9, (i)⇔(iv). Examples of $(LF)_2$- and $(LF)_3$-spaces will be given later.

The following results reveal the close interaction between $(LF)_i$-spaces, 'almost-Baire' spaces and Φ.

7.2.3 Theorem *If* $(E,\tau) = \varinjlim(E_n, \tau_n)$ *is an (LF)-space, the following statements are equivalent:*

(i) E *is an* $(LF)_3$*-space.*

(ii) E *is Baire-like.*

(iii) E *does not contain* Φ.

Proof. (iii)⇒(ii) by 6.2.3.
(ii)⇒(i) For each $n \in \mathbf{N}$, choose a countable 0-neighborhood base $(U_{n,k})_k$ in (E_n, τ_n) and set

$$\mathcal{U}_o := \{\overline{\Gamma}^E(\bigcup_{(n,k)\in H} U_{n,k}) \mid H \text{ is a finite subset of } \mathbf{N}\times\mathbf{N}\}.$$

We claim that the countable set $\mathcal{U} := \{V \in \mathcal{U}_o \mid V \text{ is a 0-neighborhood in } E\}$ is a neighborhood base of 0 in (E,τ).
Let U be a closed, absolutely convex neighborhood of 0 in E. For every $n \in \mathbf{N}$, $U \cap E_n$ is a 0-neighborhood in (E_n, τ_n), so that there exists some $k_n \in \mathbf{N}$ with $U_{n,k_n} \subseteq U \cap E_n \subseteq U$. Set $A_m := \overline{\Gamma}^E(\bigcup_{n=1}^m U_{n,k_n})$. Then $(A_m)_m$ is an absorbent sequence consisting of closed sets.
Thus $E = \bigcup_{m\in\mathbf{N}} mA_m$ and since E is Baire-like, some A_p is a 0-neighborhood in (E,τ). Clearly, $A_p \in \mathcal{U}$ and $A_p \subseteq U$. It follows that $\mathcal{U}$ is a countable neighborhood-base of 0 in (E,τ) which is therefore metrizable.
(i)⇒(iii) Φ is not metrizable. ■

7.2.4 Corollary *No (LB)-space is metrizable. In other words: there are no* $(LB)_3$*-spaces.*

Proof. It is enough to show that no (LB)-space is Baire-like.
Suppose that $(E,\tau) = \varinjlim(E_n, \tau_n)$ is a Baire-like (LB)-space. Then 6.1.4, applied to the identity $id_E : E \to E$ yields that $E \subseteq E_n$ for some $n \in \mathbf{N}$, a contradiction. ■

7.2.5 Remark The existence of $(LF)_3$-spaces (see below) shows that 6.1.4 is no longer valid if the E_n are assumed to be only Fréchet spaces. However, compare 7.1.14.

7.2.6 Theorem *If* $(E,\tau) = \varinjlim(E_n, \tau_n)$ *is an (LF)-space, the following statements are equivalent:*

(i) E is an $(LF)_2$-space.

(ii) E is quasi-Baire, but not Baire-like.

(iii) E contains Φ, but not Φ complemented.

Proof. (i)⇒(ii) E is not Baire-like by 7.2.3.
Let $(F_n)_n$ be an increasing sequence of closed subspaces covering (E, τ) and set $G_n = E_n \cap F_n$ $(n = 1, 2, ...)$. Each G_n is closed in (E_n, τ_n) (because $\tau |_{E_n} \leq \tau_n$) and $E = \bigcup_{n \in \mathbf{N}} G_n$. Set $(E, \sigma) = \underrightarrow{\lim}(G_n, \tau_n |_{G_n})$.
Now $\sigma \geq \tau$ since each $(G_n, \tau_n |_{G_n})$ is continuously included in (E, τ). By 7.1.15, $\tau = \sigma$ and $((E_n, \tau_n))_n$ and $((G_n, \tau_n |_{G_n}))_n$ are equivalent.
Since (E, τ) is of type 2 there exists a defining sequence $((H_n, \sigma_n))_n$ for (E, τ) such that each H_n is dense in (E, τ). Again by 7.1.15, $((H_n, \sigma_n))_n$and $((G_n, \tau_n |_{G_n}))_n$ are equivalent. Thus there exists some $k \in \mathbf{N}$ such that G_k is dense in (E, τ). But then $F_k = E$, implying that E is quasi-Baire.
(ii)⇒(iii) This follows from 6.2.8 and 7.2.3.
(iii)⇒(i) E is not metrizable since Φ is not. In addition to this, E is quasi-Baire by 6.2.8. Thus some E_n is not rare and therefore dense in E, so that E is of type 2. ■

7.2.7 Theorem *If $(E, \tau) = \underrightarrow{\lim}(E_n, \tau_n)$ is an (LF)-space, the following statements are equivalent:*

(i) E is an $(LF)_1$-space.

(ii) E is not quasi-Baire.

(iii) E contains Φ complemented.

(iv) E contains a closed, $\aleph_0$-codimensional subspace.

(v) E is isomorphic to $E \times \Phi$.

Proof. (ii) - (v) are equivalent by 6.2.8.
(i)⇒(ii) If E was quasi-Baire, then some E_n would be dense in E.
(ii)⇒(i) Let $(F_n)_n$ be a sequence of closed proper subspaces covering (E, τ) and define $(G_n)_n$ as in the proof of 7.2.6. Then $((G_n, \tau_n |_{G_n}))_n$ is a defining sequence for (E, τ) and $\overline{G_n} \subseteq F_n \subset E$ for each n, i.e.no G_n is dense in (E, τ). ■

7.2.8 Corollary *Every strict (LF)-space contains Φ complemented.* ■

7.2.9 Lemma *Let $(E,\tau) = \underrightarrow{\lim}(E_n, \tau_n)$, where each (E_n, τ_n) is a Fréchet space. If (E,σ) is (db) for some locally convex Hausdorff topology $\sigma \leq \tau$, then $E = E_n$ for some $n \in \mathbf{N}$ (and, by 7.1.8, $\tau = \tau_n$).*

Proof. Let $(E_n, \sigma |_{E_n})$ be dense and barrelled in (E,σ). By 7.1.8, $id_{E_n} : (E_n, \tau_n) \to (E_n, \sigma |_{E_n})$ is an isomorphism. Consequently, E_n is complete (hence closed) in (E,σ), so that $E = E_n$. ∎

7.2.9 implies that no (LF)-space is (db). Thus, no Fréchet space (being (db)) is an (LF)-space.
From this it is clear that no defining sequence $((E_n, \tau_n))_n$ for an (LF)-space (E,τ) can be eventually constant (which would imply (E,τ) to be Fréchet). Rather, any defining sequence of an (LF)-space contains a strictly increasing subsequence. So the restriction agreed upon in 7.1.3 (iii) is not substantial. Furthermore, the following result is an immediate consequence of 7.2.9:

7.2.10 Corollary *If $(E,\tau) = \underrightarrow{\lim}(E_n, \tau_n)$ is the inductive limit of a sequence of Fréchet spaces $((E_n, \tau_n))_n$, the following statements are equivalent:*

(i) (E,τ) is Fréchet.

(ii) (E,τ) is Baire.

(iii) (E,τ) is unordered Baire-like.

(iv) (E,τ) is (db).

(v) (E,τ) is not an (LF)-space, i.e. every defining sequence is eventually constant.

∎

7.2.3, 7.2.6, 7.2.7 and 7.2.10 provide large classes of distinguishig examples, to wit:

$(LF)_1$-spaces distinguish between barrelled and quasi-Baire spaces.

$(LF)_2$-spaces distinguish between quasi-Baire and Baire-like spaces.

$(LF)_3$-spaces distinguish between Baire-like and (db)-spaces.

7.2.11 Theorem *Let N be a dense, Baire-like (i.e. barrelled: see 6.2.6) subspace of a Fréchet space (F,τ). N is not a (db)-space if and only if there exists a subspace M of F and a topology σ on M such that $M \supseteq N$ and (M,σ) is an (LF)-space* *continuously included in* (F, τ).

Proof. Let $(M, \sigma) = \underrightarrow{\lim}(F_n, \sigma_n)$ be as in the hypothesis of the theorem. By 5.2.4 it is enough to show that $(M, \tau\mid_M)$ is not a (db)-space (of course, N is also dense in M).

Indeed, if $(M, \tau\mid_M)$ was (db), 7.1.9 would imply $F_n = M$ for some n, a contradiction. (This part of the proof does not use N to be Baire-like.)

Conversely, let N be a dense, barrelled, non-(db) subspace of F. Choose an increasing sequence $(G_n)_n$ of subspaces of N such that $N = \bigcup_{n=1}^{\infty} G_n$ and no G_n is both dense and barrelled in N.

Then for each $n \in \mathbf{N}$ there exists a barrel B_n in G_n whose closure $\overline{B_n}$ in F is not a neighborhood of 0 (cf. the proof of 5.2.20). Choose a countable base $(V_k)_k$ of closed neighborhoods of 0 in (F, τ) and let η_n be that locally convex topology on $H_n = sp(\overline{B_n})$ with $\{k^{-1}\overline{B_n} \cap V_k \mid k \in \mathbf{N}\}$ as a 0-neighborhood base.

(H_n, η_n) is a Fréchet space continuously included in (F, τ) by [50], I.1.6. Equip $F_k := \bigcap_{n \geq k} H_n$ with the projective topology τ_k with respect to the inclusions $F_k \hookrightarrow H_n$ $(n \geq k)$. Each (F_k, τ_k) is a Fréchet space by 7.1.12 (ii) and (F_k, τ_k) is continuously included in both (F_{k+1}, τ_{k+1}) and (F, τ) $(k = 1, 2, \ldots)$.

Suppose $N \subseteq H_n$ for some $n \in \mathbf{N}$. $\overline{B_n} \cap N$ is a neighborhood of 0 in N since N is barrelled. But then $\overline{B_n}$ would be a 0-neighborhood in F because N is dense in F (cf. 1.4.8 (ii)), a contradiction. Therefore $H_n \not\supseteq N$ for all $n \in \mathbf{N}$, so that $F_k \not\supseteq N$ for all $k \in \mathbf{N}$.

On the other hand,

$$F_k = \bigcap_{n \geq k} H_n \supseteq \bigcap_{n \geq k} sp(B_n) = \bigcap_{n \geq k} G_n = G_k,$$

implying that $N \subseteq \bigcup_{k=1}^{\infty} F_k$. Thus there exists a strictly increasing subsequence $(F_{k_j})_j$ of $(F_k)_k$ such that

$$M = \bigcup_{j=1}^{\infty} F_{k_j} = \bigcup_{k=1}^{\infty} F_k \supseteq N.$$

Now set $(M, \sigma) := \underrightarrow{\lim}(F_{k_j}, \tau_{k_j})$. Then (M, σ) is an (LF)-space continuously included in (F, τ) (since $\tau\mid_{F_{k_j}} \leq \tau_{k_j}$ for each $j \in \mathbf{N}$). ■

7.2.12 Corollary *A Fréchet space F contains a dense, Baire-like subspace N which is not (db) if and only if it contains a dense barrelled subspace M which, with a* *topology stronger than the relative topology, is an (LF)-space.*

Proof. Since N is dense and barrelled and $N \subseteq M$, the assertion follows from 5.2.3. ■

7.2.13 Proposition *Every (LF)-space $(E,\tau) = \varinjlim(E_n, \tau_n)$ possesses a defining sequence $((F_n, \sigma_n))_n$ such that no F_n is dense in (F_{n+1}, σ_{n+1}) $(n = 1, 2, \ldots)$.*

Proof. For each $n \in \mathbf{N}$, choose $x_n \in E_{n+1} \setminus E_n$ and set $(F_n, \sigma_n) := (E_n, \tau_n) \oplus sp(x_n)$. Then

$$\overline{F_n}^{\sigma_{n+1}} \subseteq \overline{E_{n+1}}^{\sigma_{n+1}} = E_{n+1} \subset F_{n+1}$$

for each n. Furthermore, the sequences $((E_n, \tau_n))_n$ and $((F_n, \sigma_n))_n$ are obviously equivalent. Hence $(E, \tau) = \varinjlim(F_n, \sigma_n)$ and the proof is complete. ■

7.2.14 Proposition *Let $(E,\tau) = \varinjlim(E_n, \tau_n)$ be an (LF)-space and suppose that E_n is dense in (E_{n+1}, τ_{n+1}) for every $n \in \mathbf{N}$. Then E_1 is dense in (E, τ). In particular, (E,τ) is of type 2 or 3.*

Proof. Since under our hypothesis E_1 is dense in each (E_{n+1}, τ_{n+1}), it follows that $\overline{E_1}^{\tau} \supseteq \overline{E_1}^{\tau_n} = E_n$ for each n and hence $\overline{E_1}^{\tau} = \bigcup_{n \in \mathbf{N}} E_n = E$. ■

7.2.15 Example *A complete $(LB)_2$-space.*

Choose $p > 1$ and $N \in \mathbf{N}$ such that $p - \frac{1}{N+1} > 1$. Endow

$$l_p^- := \bigcup_{n=1}^{\infty} l_{(p - \frac{1}{N+n})}$$

with the strongest locally convex topology τ such that every inclusion $l_{(p-\frac{1}{N+n})} \hookrightarrow l_p^-$ is continuous. As usual, let (ω, π) denote $\mathbf{K}^{\mathbf{N}}$ with the product topology.
ω is separated and $l_{(p-\frac{1}{N+n})}$ is continuously included in ω for $n = 1, 2, \ldots$, so that $\tau \geq \pi \mid_{l_p^-}$ and hence (l_p^-, τ) is separated. Thus $l_p^- = \varinjlim_{n \in \mathbf{N}} l_{(p-\frac{1}{N+n})}$ is an (LB)-space. By 7.1.4 (2), l_p^- is independent of the choice of N and by 7.2.4 and 7.2.14, l_p^- is of type 2.
Every $l_{(p-\frac{1}{N+n})}$ is a (DF)-space because it is normable (see [21], p.165). Hence by [21], p.171, Cor.1 and p.170, Cor.2, l_p^- is complete.
Obviously we also have $l_p^- = \varinjlim_{n \in \mathbf{N}} l_{p_n}$ for every strictly increasing sequence $(p_n)_n$ with $1 < p_1$ and $\lim_{n \to \infty} p_n = p$ (see 7.1.4 (2)).

In order to achieve a concise formulation of the following example we need some facts about completions of uniform spaces.

7.2.16 Lemma *Let $(X,\mathcal{U})$ and $(Y,\mathcal{V})$ be uniform Hausdorff spaces, $Z \subseteq X$, $i : Z \hookrightarrow Y$ a uniformly continuous embedding and $j : Y \to X$ uniformly continuous such that $j \circ i$ is the inclusion $Z \hookrightarrow X$.*
If $i(Z)$ is dense in Y and Z is dense in X, then j is a uniform isomorphism from Y onto a dense subspace of X. In particular, X, Y and Z have isomorphic completions.

Proof. Denote by $\widetilde{X}$ and $\widetilde{Y}$ the completions of X and Y, respectively. Since Z is dense in X, $\widetilde{X}$ is also the completion of Z. Due to the universal property of the completion of a uniform space, we obtain uniformly continuous mappings $\varphi : \widetilde{X} \to \widetilde{Y}$ and $\psi : \widetilde{Y} \to \widetilde{X}$ as follows:

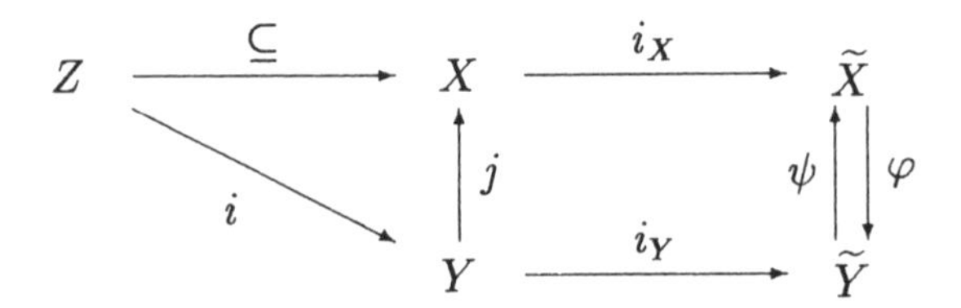

Here, $i_X : X \to \widetilde{X}$ and $i_Y : Y \to \widetilde{Y}$ are the canonical injections. It follows that $\psi \circ i_Y = i_X \circ j$ and $\varphi \circ i_X\,|_Z = i_Y \circ i$.
Now

$$\psi \circ \varphi \circ i_X\,|_Z = \psi \circ i_Y \circ i = i_X \circ j \circ i = i_X\,|_Z$$

and

$$\varphi \circ \psi \circ i_Y \circ i = \varphi \circ i_X \circ j \circ i = \varphi \circ i_X\,|_Z = i_Y \circ i.$$

Since Z (resp. $i(Z)$) is dense both in $\widetilde{X}$ and in $\widetilde{Y}$, it follows that $\psi \circ \varphi = id_{\widetilde{X}}$ and $\varphi \circ \psi = id_{\widetilde{Y}}$. Therefore j is a homeomorphic (even uniform) embedding. ■

7.2.17 Example *A metrizable (LF)-space*

With (ω, π) as in 7.2.15, set $l^1_\pi := (l^1, \pi\,|_{l^1})$. Denote by ol^1 the unit ball of the Banach space l^1 and by $\Box_k l^1$ ($k \in \mathbf{N}$) the set $\{(x_n)_n \in l^1 \mid\ |x_n| \leq 1 \text{ for } 1 \leq n \leq k\}$.
Set

$$(E, \eta) := \omega \times \omega \times \ldots \quad \text{and} \quad (F_n, \sigma_n) := \underbrace{\omega \times \omega \times \ldots \times \omega}_{n} \times l^1 \times l^1 \times \ldots$$

where σ_n is the canonical product topology ($n = 1, 2, \ldots$). $F = \bigcup_{n=1}^{\infty} F_n$ is a dense proper subspace of E and we set $(F, \sigma) := \underrightarrow{\lim}(F_n, \sigma_n)$.

Clearly $(F_n, \sigma_n) \hookrightarrow (E, \eta)$ is continuous for each n, so that $\eta \mid_F \leq \sigma$. We will show that $\sigma = \eta \mid_F$, i.e. that $(F, \eta \mid_F)$ is a dense, metrizable (LF)-subspace of the Fréchet space (E, η).

To this end, we first introduce the following definitions: For $n \in \mathbf{N}$, set

$$(G, \tau_n) := \underbrace{l^1_\pi \times ... \times l^1_\pi}_{n} \times l^1 \times l^1 \times ....$$

Then $\tau_n = \sigma_n \mid_G$ and (G, τ_n) is a dense subspace of (F_n, σ_n) for each n. (G, τ_n) is continuously included in (G, τ_{n+1}) for every n and we define $(G, \tau) = \varinjlim (G, \tau_n)$ (G is no (LF)-space since its defining sequence is not strictly increasing). Obviously, $\eta \mid_G \leq \tau$.

As a first step, we prove that $\eta \mid_G = \tau$: A typical 0-neighborhood in (G, τ) is given by $U = \Gamma(\bigcup_{n=1}^{\infty} V_n)$, where

$$V_n = \underbrace{c_n \Box_{m_n} l^1 \times ... \times c_n \Box_{m_n} l^1}_{n} \times \underbrace{c_n o l^1 \times ... \times c_n o l^1}_{m_n} \times l^1 \times l^1 \times ...,$$

where $n, m_n \in \mathbf{N}$ and $c_n > 0$.

Now set $k = m_1 + 1$ and

$$V = \underbrace{\frac{c_k}{2} \Box_{m_k} l^1 \times ... \times \frac{c_k}{2} \Box_{m_k} l^1}_{k} \times l^1 \times l^1 \times ....$$

V is an $\eta \mid_G$-neighborhood of 0 in G and $V \subseteq \Gamma(V_1 \cup V_k) \subseteq U$. Indeed, if $x = (x^{(1)}, x^{(2)}, ...) \in V$, then

$$x = (x^{(1)}, ..., x^{(k)}, 0, 0, ...) + (0, ..., 0, x^{(k+1)}, x^{(k+2)}, ...) \in \frac{1}{2} V_k + \frac{1}{2} V_1 \subseteq \Gamma(V_1 \cup V_k).$$

Thus $\tau = \eta \mid_G$. Moreover, $\sigma \mid_G = \tau$:

As noted above, $\eta \mid_F \leq \sigma$, which implies $(\tau =) \eta \mid_G \leq \sigma \mid_G$. By continuity of the inclusions $(G, \tau_n) \hookrightarrow (F_n, \sigma_n) \hookrightarrow (F, \sigma)$ (and therefore of $(G, \tau_n) \hookrightarrow (G, \sigma \mid_G)$) we infer that $\sigma \mid_G \leq \tau$. (This fact also follows from [28], 31.6 (1).)

Set $j : (F, \sigma) \hookrightarrow (E, \eta)$, $k : (G, \tau) \hookrightarrow (E, \eta)$ and $h : (G, \tau) \hookrightarrow (F, \sigma)$. We already know that h and k are topological homomorphisms and that j is continuous.

To finish the proof, it is enough to show that (E, η) together with the embedding j is isomorphic to (a copy of) the completion $(\widetilde{F}, \widetilde{\sigma})$ of (F, σ). Since $\widetilde{G} = E$ and G is dense in (F, σ) (G is dense in F_1 and F_1 is dense in (F, σ) by 7.2.14), this follows from 7.2.16.

7.2.18 Remark Since $\omega^{\mathbf{N}} \cong \omega$, 7.2.17 also shows that ω contains a dense (LF)-subspace.

7.2.19 Example *A normable (LF)-space*

Let E be the set of all continuous, bounded functions on $\mathbf{C}$ satisfying

$$p_m(f) = \int_1^\infty \mid f(\xi e^{\frac{2\pi i}{m}}) \mid d\xi < \infty$$

for $m = 1, 2, \ldots$. Define, as usual, $\|f\|_\infty := \sup_{x \in \mathbf{C}} \mid f(x) \mid$.
Denote by τ_n that locally convex topology on E with $\{\| \|_\infty\} \cup \{p_m \mid m \geq n\}$ as a generating family of seminorms. Then $(F, \tau) := \underrightarrow{\lim}(\widetilde{E_n}, \widetilde{\tau_n})$ carries the $\| \|_\infty$-topology and therefore is a normable (LF)-space : See [18], p.84 for the details.

7.3 Inheritance Properties

7.3.1 Theorem *Let $(E, \tau) = \underrightarrow{\lim}(E_n, \tau_n)$ be an (LF)-space and F a countable-codimensional subspace of E. The following statements are equivalent:*

(i) $(F, \tau \mid_F)$ is an (LF)-space .

(ii) For each $n \in \mathbf{N}$, $F \cap E_n$ is a closed subspace of (E_n, τ_n) and a proper subspace of F.

(iii) $(F, \tau \mid_F) = \underrightarrow{\lim}(F \cap E_{n_k}, \tau_{n_k} \mid_{F \cap E_{n_k}})$ for some strictly increasing subsequence $(F \cap E_{n_k})_k$ of $(F \cap E_n)_n$.

(iv) F is closed in (E, τ) and $F \not\subseteq E_n$ for each n.

Proof. (i)$\Rightarrow$(ii) Let $(F, \tau \mid_F) = \underrightarrow{\lim}(F_n, \sigma_n)$ and fix $n \in \mathbf{N}$. The (db)-space $(F \cap E_n, \tau_n \mid_{F \cap E_n})$ (cf. 5.2.11) is covered by the sequence $(F_k \cap E_n)_k$, so that some $(F_k \cap E_n, \tau_n \mid_{F_k \cap E_n})$ is barrelled and dense in $F \cap E_n$.
Denote by η the projective topology on $F_k \cap E_n$ with respect to the embeddings into (F_k, σ_k) and (E_n, τ_n). $(F_k \cap E_n, \eta)$ is a Fréchet space by 7.1.12 (ii). Thus $id : (F_k \cap E_n, \eta) \to (F_k \cap E_n, \tau_n \mid_{F_k \cap E_n})$ is an isomorphism (see 7.1.8), so that $(F_k \cap E_n, \tau_n \mid_{F_k \cap E_n})$ is a Fréchet space . Hence $F_k \cap E_n$ is a closed and dense subspace of $F \cap E_n$, i.e. $F_k \cap E_n = F \cap E_n$.
It follows that $F \cap E_n$ is closed in (E_n, τ_n) and a proper subspace of F because

$F \cap E_n = F_k \cap E_n \subseteq F_k \subset F$.

(ii)$\Rightarrow$(iii) Choose a subsequence $(n_k)_k$ of $\mathbf{N}$ such that $(F \cap E_{n_k})_k$ is strictly increasing. Set $F_k = F \cap E_{n_k}$ and $\sigma_k = \tau_{n_k} |_{F_k}$. Each (F_k, σ_k) is a Fréchet space continuously included in (F_{k+1}, σ_{k+1}) $(k = 1, 2, ...)$. Set $(F, \sigma) = \underrightarrow{\lim}(F_k, \sigma_k)$.

Since $(F_k, \sigma_k) \hookrightarrow (E, \tau)$ is continuous for every k, it follows that $\sigma \geq \tau |_F$, so that σ is separated and (F, σ) is an (LF)-space. It remains to show that $\sigma \leq \tau |_F$. Let $(L_k)_k$ be an increasing sequence of finite-dimensional subspaces such that $L = \bigcup_{n=1}^{\infty} L_n$ is an algebraic complement of F in E.

Equip $G_k := F_k + L_k$ with the unique locally convex Hausdorff topology σ_k^1 such that $\sigma_k^1 |_{F_k} = \sigma_k$ $(k = 1, 2, ...)$. Then $G_k \subset G_{k+1}$ for each k and $E = \bigcup_{k=1}^{\infty} G_k$. From $\sigma_{k+1} |_{F_k} \leq \sigma_k$ it follows that $\sigma_{k+1}^1 |_{G_k} \leq \sigma_k^1$ for each k.

Set $(E, \eta) = \underrightarrow{\lim}(G_k, \sigma_k^1)$. Now $(G_k, \sigma_k^1) \hookrightarrow (E, \tau)$ is continuous for $k \in \mathbf{N}$, so that $\eta \geq \tau$. 7.1.15 yields that $\eta = \tau$ and that $((G_k, \sigma_k^1))_k$ is equivalent to $((E_n, \tau_n))_n$. Thus for each $n \in \mathbf{N}$ there exists some k such that $E_n \subseteq G_k$ and $\sigma_k^1 |_{E_n} \leq \tau_n$.

Let V be an absolutely convex neighborhood of 0 in (F, σ) and A an absolutely convex set with $L = sp(A)$. Then

$$(V + A) \cap G_k \supseteq (V \cap G_k) + (A \cap G_k) \supseteq (V \cap F_k) + (A \cap L_k)$$

is a σ_k^1-neighborhood of 0, because $V \cap F_k$ is a σ_k-neighborhood of 0 and $A \cap L_k$ is a 0-neighborhood in L_k.

Thus, for each $n \in \mathbf{N}$, $(V+A) \cap E_n$ is a 0-neighborhood in (E_n, τ_n) since $(V+A) \cap E_n$ is a $\sigma_k^1 |_{E_n}$-neighborhood of 0 and $\sigma_k^1 |_{E_n} \leq \tau_n$. Because of the fact that, in addition to this, $(V + A)$ is absorbing in E, it is a neighborhood of 0 in (E, τ). Hence $V = (V + A) \cap F$ is a 0-neighborhood in $(F, \tau |_F)$, implying that $\sigma \leq \tau |_F$.

(iii)$\Rightarrow$(i) Clear.

(ii)$\Rightarrow$(iv) For arbitrary A and V as in (ii)$\Rightarrow$(iii), $V + A$ is a neighborhood of 0 in (E, τ). Hence $A = (V + A) \cap L$ is a 0-neighborhood in $(L, \tau |_L)$, so that τ induces on L the strongest locally convex topology.

Moreover, if $P : E = F \oplus L \to L$ denotes the projection of E onto L along F, then $p^{-1}(A) = F + A$ is a 0-neighborhood in E, so that P is continuous. It follows that F and L are topologically complementary. In particular, F is closed in (E, τ).

(iv)$\Rightarrow$(ii) This follows immediately from the fact that $\tau |_{E_n} \leq \tau_n$ for each $n \in \mathbf{N}$. ■

7.3.2 Corollary *Let F be a finite-codimensional subspace of an $(LF)_i$-space $(E, \tau) = \underrightarrow{\lim}(E_n, \tau_n)$. $(F, \tau |_F)$ is an $(LF)_j$-space if and only if it is closed and $i = j$ $(1 \leq$*

$i, j \leq 3$).

Proof. Since $(E_n)_n$ is strictly increasing, each E_n is infinite-codimensional in E and therefore $F \not\subseteq E_n$ for $n = 1, 2, \ldots$. Thus $(F, \tau|_F)$ is an (LF)-space if and only if F is closed in (E, τ).
Now suppose that F is closed and let G be an algebraic complement of F in E. Then $(E, \tau) \cong F \times G$. Hence E is Baire-like if and only if F is (see 6.3.4 and 6.3.7) and E is quasi-Baire if and only if F is (see 6.3.4 and 6.3.9). The second assertion now follows from 7.2.3, 7.2.6 and 7.2.7. ■

7.3.3 Corollary *For an (LF)-space (E, τ), the following statements are equivalent:*

(i) E is an $(LF)_1$-space.

(ii) E has an $\aleph_0$-codimensional subspace F such that $(F, \tau|_F)$ is an (LF)-space.

(iii) E has an $\aleph_0$-codimensional subspace F such that $(F, \tau|_F)$ is an (LF)-space isomorphic to E.

Proof. (i)⇒(iii) See 7.2.7, (i)⇒(v).
(iii)⇒(ii) Clear.
(ii)⇒(i) F is necessarily closed by 7.3.1, so the assertion follows from 7.2.7. ■

During the following considerations, always keep in mind 7.1.20 (iii).

7.3.4 Proposition *If $(E, \sigma) = \varinjlim(E_n, \sigma_n)$ and $(F, \tau) = \varinjlim(F_n, \tau_n)$ are (LF)-spaces (resp. (LB)-spaces), then $(E \oplus F, \sigma \oplus \tau)$ is an (LF)-space (resp. (LB)-space) with defining sequence $((E_n \oplus F_n, \sigma_n \oplus \tau_n))_n$.*

Proof. Since $(E_n)_n$ and $(F_n)_n$ are increasing, $E \oplus F = \bigcup_{n=1}^{\infty}(E_n \oplus F_n)$. Clearly, for $n \in \mathbf{N}$ we have $E_n \oplus F_n \subset E_{n+1} \oplus F_{n+1}$ and

$$(\sigma_{n+1} \oplus \tau_{n+1})|_{E_n \oplus F_n} = \sigma_{n+1}|_{E_n} \oplus \tau_{n+1}|_{F_n} \leq \sigma_n \oplus \tau_n.$$

Now $\sigma \oplus \tau$ is the finest locally convex topology on $E \oplus F$ such that $E \hookrightarrow E \oplus F$ and $F \hookrightarrow E \oplus F$ (equivalently, all $E_n \hookrightarrow E \oplus F$ and all $F_n \hookrightarrow E \oplus F$) are continuous. The topology of $\varinjlim(E_n \oplus F_n)$ is the finest locally convex topology on $E \oplus F$ such that all embeddings $E_n \oplus F_n \hookrightarrow E \oplus F$ (equivalently, all $E_n \hookrightarrow E \oplus F$ and all $F_n \hookrightarrow E \oplus F$) are continuous. Therefore, these two topologies on $E \oplus F$ coincide. ■

An obvious modification of the proof of 7.3.4 yields

7.3.5 Proposition *If $(E,\sigma)=\varinjlim(E_n,\sigma_n)$ is an (LF)-space (resp. (LB)-space) and (F,τ) is a Fréchet (resp. Banach) space, then $(E\oplus F,\sigma\oplus\tau)$ is an (LF)-space (resp. (LB)-space) with defining sequence $((E_n\oplus F,\sigma_n\oplus\tau))_n$.* ■

7.3.6 Theorem *The product of an $(LF)_i$-space (resp. $(LB)_i$-space) E with an $(LF)_j$-space (resp. $(LB)_j$-space) F is an $(LF)_k$-space (resp. $(LB)_k$-space), where $k=\min(i,j)$ $(1\le i,j\le 3)$.*

Proof. $E\times F$ is an (LF)-space (resp. (LB)-space) by 7.3.4 (cf. 7.1.20 (iii)).
If E is of type 1, then E is not quasi-Baire (7.2.7). Thus $E\times F$ is not quasi-Baire (apply 6.3.5 (i) to $\pi_1: E\times F\to E$) and hence an $(LF)_1$-space.
If both E and F are of type 2 then they are quasi-Baire but not Baire-like by 7.2.6. Then $E\times F$ is quasi-Baire (6.3.10) but not Baire-like (6.3.5 (iv), applied to $\pi_1: E\times F\to E$) and consequently it is an $(LF)_2$-space. The same argument yields that $E\times F$ is of type 2 if E is $(LF)_2$ and F is $(LF)_3$.
Finally, if both E and F are of type 3 then they are Baire-like (7.2.3). Thus $E\times F$ is Baire-like (6.3.7) and therefore of type 3. ■

7.3.7 Remark The proof of 7.3.6 shows that if E is an $(LF)_i$-space (resp. $(LB)_i$-space) and F is a Fréchet (resp. Banach) space, then $E\times F$ is an $(LF)_i$-space (resp. $(LB)_i$-space) $(1\le i\le 3)$.

7.3.8 Lemma *No (LF)-space is the continuous linear image of a Fréchet space.*

Proof. Assume there exists a continuous linear surjection $u: E\to F$, E Fréchet, F an (LF)-space. $E/f^{-1}(0)$ again is Fréchet while F is barrelled. By 7.1.8, F is topologically isomorphic to $E/f^{-1}(0)$ which is impossible. ■

There is, however, no possibility of generalizing 7.3.6 to infinite products:

7.3.9 Theorem *If $(E,\sigma)=\prod_{i\in I}(E^{(i)},\sigma^{(i)})$ is an infinite product of (LF)-spaces $(E^{(i)},\sigma^{(i)})=\varinjlim(E_n^{(i)},\sigma_n^{(i)})$, then (E,σ) is not an (LF)-space.*

Proof. Assume that $(E,\sigma)=\varinjlim(F_n,\tau_n)$ is an (LF)-space. As usual, if $J\subseteq I$ and for $i\in I$ L_i is a subspace of $E^{(i)}$ we will identify $\prod_{i\in I}L_i$ as a subspace of E.
Suppose there exist $i\in I$ and $n\in\mathbf{N}$ with $E^{(i)}\subseteq F_n$. Then $E^{(i)}$ is closed in (F_n,τ_n) (since $\sigma\,|_{F_n}\le\tau_n$) and

$$\sigma^{(i)}=\sigma\,|_{E^{(i)}}=\sigma\,|_{F_n}|_{E^{(i)}}\le\tau_n\,|_{E^{(i)}}\ .$$

But then $id : (E^{(i)}, \tau_n |_{E^{(i)}}) \to (E^{(i)}, \sigma^{(i)})$ is a continuous linear surjection from a Fréchet space onto an (LF)-space, contradicting 7.3.8.Therefore $E^{(i)} \not\subseteq F_n$ for all $i \in I$ and all $n \in \mathbf{N}$.

Let $i_1, i_2, \ldots$ be distinct members of I. For each $n \in \mathbf{N}$ we can choose $k_n \in \mathbf{N}$ such that $E_{k_n}^{(i_n)} \not\subseteq F_n$. Then $G := \prod_{n=1}^{\infty} E_{k_n}^{(i_n)}$ is a Fréchet space continuously included in E but not in any F_n. However, this contradicts 7.1.14 and the proof is complete. ■

Not surprisingly, (LF)-spaces show a much more satisfying behaviour in their interaction with other inductive limits:

7.3.10 Theorem *Let (E, τ) be the inductive limit of an increasing sequence of (LF)-spaces $(E^{(i)}, \tau^{(i)}) = \underrightarrow{\lim}(E_n^{(i)}, \tau_n^{(i)})$, $i = 1, 2, \ldots$ such that $(E^{(i)}, \tau^{(i)})$ is continuously included in $(E^{(i+1)}, \tau^{(i+1)})$ for each $i \in \mathbf{N}$. If τ is Hausdorff, then (E, τ) is an (LF)-space.*

Proof. Since $((E^{(i)}, \tau^{(i)}))_i$ is increasing, a repeated application of 7.1.14 yields the existence of a strictly increasing sequence $(m_k)_k$ of natural numbers ($m_1 = 1$) such that $E_{m_{k+1}}^{(k+1)}$ properly and continuously includes $E_p^{(i)}$ for $1 \leq i \leq k$ and $1 \leq p \leq m_k$ ($k \in \mathbf{N}$).

Set $(E, \sigma) = \underrightarrow{\lim}(E_{m_k}^{(k)}, \tau_{m_k}^{(k)})$. (Clearly, $E = \bigcup_{k=1}^{\infty} E_{m_k}^{(k)}$.) Now σ is the inductive topology with respect to the inclusions $(E_{m_k}^{(k)}, \tau_{m_k}^{(k)}) \hookrightarrow E$, hence also with respect to all inclusions $(E_n^{(i)}, \tau_n^{(i)}) \hookrightarrow E$.

The continuity of all these mappings, however, is equivalent to the continuity of all inclusions $(\lim_n E_n^{(i)}, \tau^{(i)}) \hookrightarrow E$, for which τ is the respective inductive topology. Thus $\sigma = \tau$. ■

Next we want to examine some basic properties of quotients of (LF)-spaces. The results achieved here will prove most valuable in tackling the problem of constructing metrizable and normable (LF)-spaces.

7.3.11 Theorem *Let M be a closed subspace of the $(LF)_i$-space $(E, \sigma) = \underrightarrow{\lim}(E_n, \sigma_n)$ ($1 \leq i \leq 3$). If $E_n + M = E$ for some n, then E/M is a Fréchet space. Otherwise, E/M is an $(LF)_j$-space for some $j \geq i$.*

Proof. 1.) Let $E_n + M = E$ for some n and denote by $\pi : E \to E/M$ the canonical projection. Then

$$\varphi : (E_n, \sigma_n) \hookrightarrow (E, \sigma) \xrightarrow{\pi} E/M$$

is a continuous linear surjection from the Fréchet space (E_n, σ_n) onto the barrelled space E/M. Thus $\widehat{\varphi} : E_n/\varphi^{-1}(0) \to E/M$ is an isomorphism by 7.1.8, so that E/M is a Fréchet space.

2.) Now suppose that $E_n + M \subset E$ for each $n \in \mathbf{N}$. Then $\pi(E_n) \subset E/M$ for $n = 1, 2, \ldots$. By passing to a suitable subsequence of $((E_n, \sigma_n))_n$ (which we again denote by $((E_n, \sigma_n))_n$) we can assume that $(\pi(E_n))_n$ is strictly increasing.

For $n \leq m$, consider the following diagram:

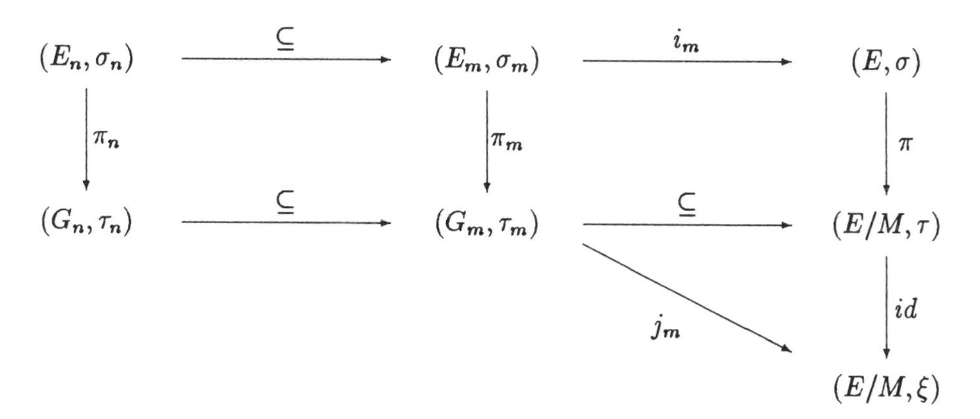

Here, $G_n = E_n/(E_n \cap M)$, $G_m = E_m/(E_m \cap M)$. π_n, π_m and π are the quotient mappings and τ_n, τ_m, τ the respective quotient topologies. The upper horizontal arrows in the above picture denote the respective canonical embeddings.

Due to the factorizing property of quotients, the lower horizontal arrows exist as continuous linear maps which, moreover, are injective since

$$E_n/(E_n \cap M) \cong (E_n + M)/M \cong \pi(E_n)$$

algebraically. Set $(E/M, \xi) := \underrightarrow{\lim}(G_n, \tau_n)$.

τ is the inductive topology with respect to π and ξ is the inductive topology with respect to the family $(j_m)_m$. Now π is continuous iff each $\pi \circ i_m = j_m \circ \pi_m$ is continuous iff each j_m is continuous. Thus $\pi = \xi$. This implies that E/M is an $(LF)_j$-space for some $1 \leq j \leq 3$.

If some E_n is dense in E, then

$$E/M = \pi(E) = \pi(\overline{E_n}) \subseteq \overline{\pi(E_n)},$$

so that G_n is dense in E/M. Moreover, E/M is metrizable whenever E is, so that $j \geq i$. ∎

As a direct consequence, we get an alternative proof of 7.1.4 (8):

7.3.12 Corollary *Every continuous linear surjection f from an (LF)-space (E,σ) onto an (LF)-space (F,τ) is open.*

Proof. Set $M := f^{-1}(0)$. We have to show that the associated injection (which, in fact, is a bijection) $\widehat{f} : E/M \to F$ is an isomorphism.
If E/M were a Fréchet space then so would be F by 7.1.8, a contradiction. Hence E/M, equipped with the quotient topology η is an (LF)-space by 7.3.11. Identifying E/M with F algebraically, we have $\eta \geq \tau$ for the two (LF)-topologies η, τ on F. Therefore $\eta = \tau$ by 7.1.16, i.e. $\widehat{f}$ is an isomorphism. ∎

7.3.13 Theorem *Let $(E,\sigma) = \varinjlim(E_n,\sigma_n)$ be an $(LF)_i$-space and M a closed subspace of (E,σ) contained in E_1. Then $(F,\tau) = (E,\sigma)/M$ is an $(LF)_j$-space for some $j \geq i$ with defining sequence $((E_n,\sigma_n)/M)_n$.*

Proof. Since $M + E_n \subseteq E_1 + E_n = E_n \subset E$ for each $n \in \mathbf{N}$, this follows from the proof of 7.3.11. ∎

7.3.14 Corollary *If $(E,\sigma) = \varinjlim(E_n,\sigma_n)$ is a metrizable (LF)-space and M is a complete subspace of (E,σ), then $(E,\sigma)/M$ is a metrizable (LF)-space.*

Proof. Since $(M, \sigma\mid_M)$ is a Fréchet space, it is (even continuously) included in some (E_{n_o}, σ_{n_o}) by 7.1.14. Now 7.3.13, applied to the defining sequence $((E_n,\sigma_n))_{n\geq n_o}$, yields the result. ∎

7.3.15 Theorem *Let f be a continuous linear surjection from a Fréchet space (F,σ) onto a Fréchet space (G,τ). The following statements are equivalent:*

(i) G has a dense subspace G_o such that $(G_o, \tau\mid_{G_o})$ is an (LF)-space.

(ii) F has a dense subspace F_o such that $(F_o, \sigma\mid_{F_o})$ is an (LF)-space containing $f^{-1}(0)$.

Proof. (i)⇒(ii) Let $(G_o, \tau\mid_{G_o}) = \varinjlim(G_n,\tau_n)$. For each $n \in \mathbf{N}$, endow $F_n := f^{-1}(G_n)$ with σ_n, the projective topology with respect to the embedding $F_n \hookrightarrow (F,\sigma)$ and the map $f\mid_{F_n}: F_n \to G_n$.
By 7.1.12 (i), $((F_n,\sigma_n))_n$ is a strictly increasing sequence of Fréchet spaces and we set $(F_o,\sigma_o) := \varinjlim(F_n,\sigma_n)$. Then

$$F_o = \bigcup_{n=1}^{\infty} F_n = \bigcup_{n=1}^{\infty} f^{-1}(G_n) = f^{-1}(\bigcup_{n=1}^{\infty} G_n) = f^{-1}(G_o) \supseteq f^{-1}(0).$$

f is open by 7.1.8. Hence if $\emptyset \neq U$ is open in F then $f(U) \cap G_o \neq \emptyset$, so that $U \cap f^{-1}(G_o) \neq \emptyset$, implying that F_o is dense in F.
Since for each n (F_n, σ_n) is continuously included in (F, σ), hence in $(F_o, \sigma |_{F_o})$, we have $\sigma |_{F_o} \leq \sigma_o$.
Conversely, let W be an absolutely convex σ_o-neighborhood of 0 in F_o. Then $W \cap F_n$ is a σ_n-neighborhood of 0 in F_n for each n, so that we can find absolutely convex 0-neighborhoods U_n and V_n in (F, σ) and (G_n, τ_n), respectively, with $U_n \cap f^{-1}(V_n) \subseteq W \cap F_n$. Hence

$$f(W) \supseteq f(U_n \cap f^{-1}(V_n)) = f(U_n) \cap V_n,$$

so that $f(W) \cap G_n \supseteq f(U_n) \cap V_n$ is a τ_n-neighborhood of 0 for each $n \in \mathbf{N}$ since f is open and $\tau |_{G_n} \leq \tau_n$. This implies that $f(W) = f(W) \cap G_o$ is a 0-neighborhood in the inductive limit topology $\tau |_{G_o}$. In other words, we have shown that $f |_{F_o}: (F_o, \sigma_o) \to (G_o, \tau |_{G_o})$ is open.
Since $f |_{F_o}$ (now considered as a map from $(F_o, \sigma |_{F_o})$ to $(G_o, \tau |_{G_o})$ again) is continuous, $f^{-1}(f(W)) = f^{-1}(0) + W$ is a $\sigma |_{F_o}$-neighborhood of 0 in F_o.
Unfortunately, $f^{-1}(f(W))$ is, in general, not a subset of the σ_o-0-neighborhood W, so we are not yet done. However, as we will show, the smaller set

$$W_o := \frac{1}{3}(U_1 \cap f^{-1}(f(W \cap U_1)))$$

is contained in W yet still is a $\sigma |_{F_o}$-neighborhood of 0 which will establish $\sigma |_{F_o} = \sigma_o$.
First of all, $3W_o = U_1 \cap (f^{-1}(0) + (W \cap U_1))$. Let $y \in U_1 \cap (f^{-1}(0) + (W \cap U_1))$. Then $y = z + w$, $z \in f^{-1}(0)$ and $w \in W \cap U_1$. Thus $z = y - w \in U_1 - U_1 = 2U_1$, so that $z \in 2U_1 \cap f^{-1}(0)$ and

$$y = z + w \in (2U_1 \cap 2f^{-1}(0)) + W \subseteq 2(U_1 \cap f^{-1}(V_1)) + W \subseteq 2W + W = 3W.$$

Hence, indeed, $W_o \subseteq W$.
On the other hand, $U_1 \cap F_o$ is an absolutely convex 0-neighborhood in F_o with respect to $\sigma |_{F_o}$, hence with respect to σ_o. So $W \cap (U_1 \cap F_o) = W \cap U_1$ can take the place of W in the argument preceding the introduction of W_o. We obtain that $f^{-1}(f(W \cap U_1))$ (hence also W_o) is a 0-neighborhood with respect to $\sigma |_{F_o}$.
(ii)$\Rightarrow$(i) Set $M = f^{-1}(0)$ and $G_o = f(F_o)$. Since M is a subspace of F_o, F_o/M is a topological subspace of F/M by 3.1.5. Now $\widehat{f} : F/M \to G$, $\widehat{f}(x + M) = f(x)$ is an isomorphism by 7.1.8, so that $F_o/M \cong (G_o, \tau |_{G_o})$.

Moreover, F_o/M is an (LF)-space because of 7.3.14. Finally,

$$G = f(\overline{F_o}) \subseteq \overline{f(F_o)} = \overline{G_o},$$

i.e. G_o is dense in G. ■

7.3.16 Example *A complete, non strict $(LB)_1$-space.*

Set $E = \Phi \times l_p^-$, where $l_p^- = \underset{\longrightarrow}{\lim} l_{p_n}$ is defined in 7.2.15. Let $(G_n)_n$ be a defining sequence for Φ (each G_n is necessarily finite-dimensional, cf. 7.1.21). E is an $(LB)_1$-space by 7.3.6. E is complete since Φ and l_p^- are complete.
Suppose $E = \underset{\longrightarrow}{\lim}(E_k, \sigma_k)$ is strict.
By 7.1.15, there exist $k_1, ..., k_5$ such that each of the following inclusions is continuous:

$$l_{p_1} \times G_1 \hookrightarrow E_{k_1} \hookrightarrow l_{p_{k_2}} \times G_{k_2} \hookrightarrow E_{k_3} \hookrightarrow l_{p_{k_4}} \times G_{k_4} \hookrightarrow E_{k_5}.$$

Denote by τ_n the topology on $l_{p_n} \times G_n$ for $n \in \mathbf{N}$. Since σ_{k_5} induces σ_{k_1} on E_{k_1}, so do τ_{k_2}, σ_{k_3} and τ_{k_4}:

$$\sigma_{k_5} \mid_{E_{k_1}} \leq \tau_{k_4} \mid_{E_{k_1}} \leq \sigma_{k_3} \mid_{E_{k_1}} \leq \tau_{k_2} \mid_{E_{k_1}} \leq \sigma_{k_1} = \sigma_{k_5} \mid_{E_{k_1}} .$$

Hence $l_{p_{k_2}} \times G_{k_2}$ and $l_{p_{k_4}} \times G_{k_4}$ induce the same topology on $l_{p_1} \times G_1$, implying that $l_{p_{k_2}}$ and $l_{p_{k_4}}$ induce the same topology on l_{p_1}.
But this is a contradiction because $p_1 < p_{k_2} < p_{k_4}$.

7.4 The Separable Quotient Problem

It is a long standing open question wether every infinite-dimensional Banach space E possesses an infinite-dimensional separable quotient by a closed subspace or, equivalently, if every infinite-dimensional Banach space can be mapped linearly and continuously onto an infinite-dimensional separable Banach space.
Up to now, there are only partial results in the form of general theorems stating that each Banach space from a certain class has a separable quotient. For example, an affirmative answer to the above question is given in [31] and [43] if $E = C(X)$, X a compact Hausdorff space (i.e. if E is an abelian C^*-Algebra).
This section is intended to point out the surprisingly close relationship between the separable quotient problem in Fréchet and Banach spaces and the existence of metrizable and normable (LF)-spaces.

Unless otherwise stated, by a separable quotient we will always mean an infinite-dimensional separable quotient by a closed subspace. First of all, we need some preparations concerning separable quotients of non-Banach Fréchet spaces. For this purpose, we make use of a classical paper by M.EIDELHEIT ([17]):

7.4.1 Definition *Let (E, τ) be a Fréchet space and $(p_k)_{k\in\mathbf{N}}$ a generating sequence of seminorms for τ. If $0 \neq f \in E'$, then the* order $n(f)$ *of f with respect to $(E, (p_k)_k)$ is the smallest natural number N for which there exists some $M > 0$ such that*

$$| f(x) | \leq M \sum_{i=1}^{N} p_i(x)$$

for each $x \in E$. We set $n(f) = 0$ for $f \equiv 0$.

7.4.2 Proposition *Let $(E, (p_k)_k)$ be as in 7.4.1 and suppose there exists a sequence $(f_i)_i$ in E' such that $n(f_i) < n(f_{i+1})$ for each $i \in \mathbf{N}$. Then $f : x \to (f_i(x))_{i\in\mathbf{N}}$ is a continuous linear surjection from E onto ω ($= \mathbf{K}^{\mathbf{N}}$ with the product topology). In particular, E has a quotient isomorphic to ω.*

Proof. See [17], p. 143/144, Satz 2 and Bemerkung 3. ■

7.4.3 Remark There is an interesting application of 7.4.2 to real analysis: Consider the (Fréchet-) subspace

$$E := \{f \in \mathcal{D}(\mathbf{R}) \mid supp(f) \subseteq [-1, 1]\}$$

of $\mathcal{D}(\mathbf{R})$ (cf. 7.1.6) and set $f_n : E \to \mathbf{R}$, $\varphi \to \varphi^{(n)}(0)$ for $n \in \mathbf{N}$.
The sequence $(f_n)_n$ satisfies the hypothesis of 7.4.2, thereby establishing the fact that for each real sequence $(\alpha_k)_{k=0}^{\infty}$ there exists an infinitely differentiable real valued function φ on $\mathbf{R}$ having compact support such that $\varphi^{(n)}(0) = \alpha_n$ for $n \geq 0$.
Observe that for $\limsup \sqrt[n]{| \alpha_n |} = \infty$, the Taylor series

$$\varphi(x) := \sum_{k=0}^{\infty} \frac{\alpha_k}{k!} x^k$$

does not work even if non-compact support is admitted.

7.4.4 Corollary *Every non-Banach Fréchet space has a quotient isomorphic to ω.*

Proof. Let $(p_k)_k$ be a generating sequence of seminorms for the topology τ on E. By 7.4.2 we have to prove the existence of a sequence $(f_i)_i$ in E' such that

$n(f_i) < n(f_{i+1})$ for each $i \in \mathbf{N}$.
Suppose no such sequence exists and choose $N \in \mathbf{N}$ such that $n(f) \leq N$ for all $f \in E'$. Then for every $f \in E'$ there exists some $M \in \mathbf{N}$ such that

$$| f(x) | \leq M \sum_{i=1}^{N} p_i(x) \quad \text{for each } x \in E. \tag{$*$}$$

Denote by τ_N that locally convex topology on E defined by $p = \max_{1 \leq i \leq N} p_i$. Then $\tau_N \leq \tau$ and $(E, \tau)' = (E, \tau_N)'$ by $(*)$. Again by $(*)$, if $p(x) = 0$ then $f(x) = 0$ for each $f \in E'$, so that $x = 0$ by the Hahn-Banach theorem. Hence p is a norm and therefore (E, τ_N) is a Mackey space. But then $(E, \tau) = (E, \tau_N)$ is a Banach space, a contradiction. ■

For an independent proof of 7.4.4, cf. [5], 2.6.16.

7.4.5 Remark Since ω is a separable Fréchet space, 7.4.4 provides an affirmative solution to the separable quotient problem for non-Banach Fréchet spaces.

7.4.6 Proposition *Let F be a dense, non-barrelled subspace of a Banach space E. Then there exists an increasing sequence $(S_n)_n$ of closed subspaces of E such that $\bigcup_{n=1}^{\infty} S_n$ is dense in E and $dim(S_n/S_{n-1}) = 1$ for each n.*

Proof. Choose a barrel in F which is not a neighborhood of 0 in F and set $B_1 := \overline{B}$. B_1 is a barrel in $sp(B_1)$, yet no 0-neighborhood: If it were, $B = B_1 \cap F$ would be a 0-neighborhood in F. Therefore, $sp(B_1)$ is not barrelled either. By 1.1.1 it has infinite codimension in E. $sp(B_1)$ is dense because it contains F.
We will inductively define sequences $(x_n)_n$ in E, $(f_n)_n$ in E' and closed, absolutely convex sets $B_n := B_1 + \{\sum_{i=1}^{n-1} a_i x_i \mid \ | a_i | \leq 1\}$ such that for $n \in \mathbf{N}$

(i) $\|x_n\| = 1$, $x_n \notin sp(B_n)$ (in particular, $x_n \notin 2^n B_n$).

(ii) $f_i(x_n) = 0$ for $i < n$, $f_n(x_n) = 1$, $| f_n | \leq \frac{1}{2^n}$ on B_n.

We begin by selecting x_1 and f_1: Since $sp(B_1)$ has infinite codimension, there exists $x_1 \notin sp(B_1)$ satisfying $\|x_1\| = 1$. By the Hahn-Banach theorem there is $f_1 \in E'$ such that $f_1(x_1) = 1$ and $| f_1 | \leq \frac{1}{2}$ on B_1.
Now assume that for some $n \geq 2$, $x_1, ..., x_{n-1}$ and $f_1, ..., f_{n-1}$ have been chosen to satisfy (i) and (ii). B_n is absolutely convex and closed as the sum of a closed and a compact set. $sp(B_1)$ has codimension at most (in fact, equal to) n in $sp(B_n)$. Therefore, $sp(B_n)$ has infinite codimension in E. It follows that $\bigcap_{i=1}^{n-1} f_i^{-1}(0)$, being

of finite codimension in E, cannot be a subset of $sp(B_n)$.
Thus there exists $x_n \in \bigcap_{i=1}^{n-1} f_i^{-1}(0) \setminus sp(B_n)$ such that $\|x_n\| = 1$. Again by the Hahn-Banach theorem there exists $f_n \in E'$ satisfying $f_n(x_n) = 1$ and $| f_n | \leq \frac{1}{2^n}$ on B_n, so that the induction step is complete. By (ii) we have $| f_i(x_n) | \leq \frac{1}{2^i}$ for $i > n$.
Fix $x \in B_1$, $k \in \mathbf{N}$ and set $a_1 := -f_{k+1}(x)$. Then $| a_1 | \leq \frac{1}{2^{k+1}}$. Set

$$a_n = -f_{k+n}(\underbrace{x + \sum_{i=1}^{n-1} a_i x_{k+i}}_{\in B_{k+n}}).$$

Then $| a_n | \leq \frac{1}{2^{k+n}}$ by (ii). For $z = \sum_{i=1}^{\infty} a_i x_{k+i}$ we have $\|z\| = 1 \leq \frac{1}{2^k}$.
Moreover,

$$f_{k+n}(x+z) = f_{k+n}(x + \sum_{i=1}^{n-1} a_i x_{k+i}) + a_n + \sum_{i=n+1}^{\infty} a_i f_{k+n}(x_{k+i}) = -a_n + a_n + 0 = 0.$$

Now define $S_k := \bigcap_{i=k+1}^{\infty} f_i^{-1}(0)$ and $S = \bigcup_{k=1}^{\infty} S_k$. Then $d(x, S_k) \leq \frac{1}{2^k}$, so that $d(x, S) = 0$. It follows that S is dense in $sp(B_1)$ and therefore in E.
It remains to show that indeed $dim(S_k/S_{k-1}) = 1$ for each k. $S_{k-1} = S_k \cap f_k^{-1}(0)$, so $dim(S_k/S_{k-1}) \leq 1$. We will prove that $S_k \setminus S_{k-1} \neq \emptyset$, thereby establishing our claim.
To this end, we replace $x \in B_1$ by x_k in the above construction of the scalars a_i and $z \in E$. The induction remains valid since $x_k \in B_{k+1}$ and also $x_k + \sum_{i=1}^{n-1} a_i x_{k+i} \in B_{k+n}$ for $| a_i | \leq 1$.
It follows that $f_{k+n}(x_k + z) = 0$ for $n \in \mathbf{N}$ as before, yet

$$f_k(x_k + z) = f_k(x_k) + \sum_{i=1}^{\infty} a_i f_k(x_{k+i}) = 1.$$

Therefore $x_k + z \in S_k \setminus S_{k-1}$. ■

We are now in a position to give the following reformulations of the separable quotient problem:

7.4.7 Theorem *For a Fréchet space (E, τ), the following statements are equivalent:*

(i) E has a separable, infinite-dimensional quotient by a closed subspace.

(ii) There exists a strictly increasing sequence $(F_n)_n$ of closed subspaces of E such that $F = \bigcup_{n=1}^{\infty} F_n$ is dense in E (in short, F is a dense S_σ-subspace of E).

(iii) E has a dense, non-barrelled subspace F.

(iv) E has a dense, non-(db)-subspace F.

(v) E densely, properly and continuously includes a Fréchet space (G, σ).

Proof. (i)⇒(ii) Let M be a closed subspace of E such that E/M is infinite-dimensional and separable and choose a sequence $(x_n+M)_n$ in E/M which is linearly independent and spans a dense subspace. Then $(x_n)_n$ is linearly independent in E, $M \cap sp(\{x_n \mid n \in \mathbf{N}\}) = \{0\}$ and $M + sp(\{x_n \mid n \in \mathbf{N}\})$ is dense in E. Now set

$$F := \bigcup_{n=1}^{\infty}(M + sp(\{x_1, ..., x_n\})).$$

(ii)⇒(iii) Let F be as in (ii) and suppose that F is barrelled. By 6.2.6, F is Baire-like, so that some $F_n = F$, contradicting the fact that $(F_n)_n$ is strictly increasing.
(iii)⇒(i) If E is non-Banach, (i) holds by 7.4.4. Now let E be a Banach space, choose $(S_n)_n$ as in 7.4.6 and set $S_n = S_{n-1} \oplus sp(y_n)$ for $n \geq 2$. Denote by $\pi : E \to E/S_1$ the canonical projection. Since $\bigcup_{n=1}^{\infty} S_n = S_1 + sp(\{y_n \mid n \in \mathbf{N}\})$ is dense in E, $\pi(S_1 + sp(\{y_n \mid n \in \mathbf{N}\})) = \pi(sp(\{y_n \mid n \in \mathbf{N}\}))$ is dense in E/S_1. Since $sp(\{y_n \mid n \in \mathbf{N}\})$ is separable, (i) follows.
(iii)⇒(iv) Obvious.
(iv)⇒(iii) F is the union of an increasing sequence $(F_n)_n$ of subspaces, none of which is both dense and barrelled in F. If F is not barrelled, (iii) is clearly satisfied.
If F is barrelled then it is Baire-like and therefore quasi-Baire by 6.2.6. Thus some F_n is dense in F and therefore dense and non-barrelled in E.
(iii)⇒(v) Let V_1 be a barrel in F which is not a neighborhood of 0 in F and set $V := \overline{V_1}$. As in the proof of 7.4.6, V is a barrel in $G := sp(V)$ but not a 0-neighborhood in G. Consequently, G is a proper subspace of the (barrelled) space E.
Let σ be that locally convex topology on G having

$$\{k^{-1}V \cap U \mid U \text{ is a closed neighborhood of 0 in } (E, \tau), k \in \mathbf{N}\}$$

as a neighborhood base of the origin. Then (G, σ) is complete (cf. [50], I.1.6), metrizable and continuously included in (E, τ).
G is dense in E since $F \subseteq G$ and proper as noted above.
(v)⇒(iii) Suppose that $(G, \tau|_G)$ is barrelled. By 7.1.8, $id : (G, \sigma) \to (G, \tau|_G)$

is an isomorphism. But then G is dense and complete in (E, τ), i.e. $G = E$, a contradiction. ■

7.4.8 Remarks

(i) By 7.4.5, every non-Banach Fréchet space satisfies (i)-(v) of 7.4.7.

(ii) If E is a Banach space, (i)-(v) of 7.4.7 are equivalent to the following conditions:

(a) E densely, properly and continuously includes a Banach space F.

(b) There exists a sequence $(f_n)_n$ in E' such that $\|f_n\| = 1$ for all n and $\bigcup_{n=1}^{\infty} \bigcap_{k=n}^{\infty} f_k^{-1}(0)$ is dense in E.

(c) There exist closed subspaces F and G of E such that $F + G$ is dense in and proper in E.

For the proofs of these results, see [49].

(iii) 7.5.5 and 7.5.7 will provide further equivalences that directly demonstrate the close relationship between $(LF)_3$-spaces and the separable quotient problem.

7.4.9 Lemma *Let M_1 and M_2 be linear subspaces of a tvs (E, τ) with $M_1 \subseteq M_2 \subseteq E$. Denote by π_1, π_2 the canonical projections onto the quotients $(E/M_1, \tau_1)$ resp. $(E/M_2, \tau_2)$. Then the induced map $\pi : E/M_1 \to E/M_2$ is, in its turn, a quotient map, $(E/M_2, \tau_2)$ being isomorphic to $((E/M_1)/\pi_1(M_2), \tau_3)$:*

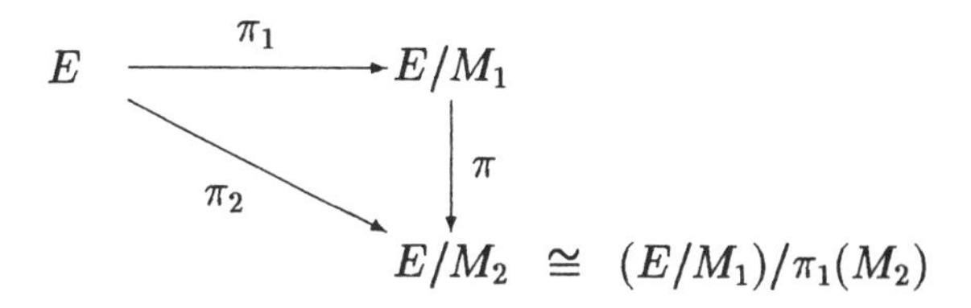

Moreover, M_2 is closed in (E, τ) if and only if $\pi_1(M_2)$ is closed in $(E/M_1, \tau_1)$.

Proof. Due to the factorization property of $(E/M_1, \pi_1)$, π exists as continuous linear map and obviously is surjective. Hence, algebraically we have $E/M_2 = (E/M_1)/ker(\pi)$, where

$$ker(\pi) = \{\pi_1(x) \mid x \in E, \pi(\pi_1(x)) = \pi_2(x) = 0\} = \pi_1(ker(\pi_2)) = \pi_1(M_2).$$

τ_2 and τ_3 are defined as the inductive topologies with respect to π_2 resp. π. Now π is continuous with respect to some locally convex topology on E/M_2 if and only if $\pi \circ \pi_1 = \pi_2$ is so. Therefore, $\tau_2 = \tau_3$.
Finally, $\pi_1^{-1}(\pi_1(M_2)) = M_2 + M_1 = M_2$. By definition of the quotient topology, $\pi_1(M_2)$ is τ_1-closed if and only if M_2 is τ-closed. (Alternatively, observe that each of the two conditions is equivalent to $\tau_2 = \tau_3$ being Hausdorff.) ■

7.4.10 Lemma *Let (F,τ) be an lcs and (L,σ) an lcs of finite dimension. $F \oplus L$ has a separable (infinite-dimensional) Hausdorff quotient if and only if F has.*

Proof. First observe that F is a quotient of $F \oplus L$. If F has a separable quotient F/M, then the same is true of $F \oplus L$ by 7.4.9: $(F \oplus L)/(M \oplus L)$ is isomorphic to F/M (consider $L \subseteq M \oplus L \subseteq F \oplus L$).
Conversely, assume that M is a closed subspace of $F \oplus L$ such that $(F \oplus L)/M$ is separable and has infinite dimension. Then $M + L$ is also closed in $F \oplus L$.
Twofold application of 7.4.9 yields the upper and lower triangle of the following diagram (containing only quotient maps):

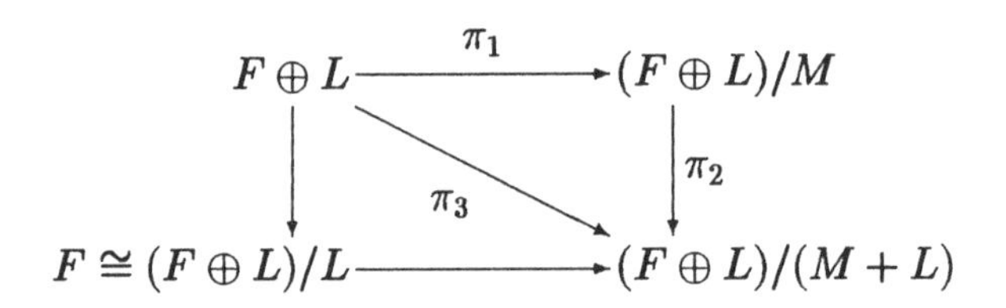

Now $ker(\pi_2) = \pi_1(M+L) = \pi_1(L)$ is of finite dimension. Therefore $(F \oplus L)/(M+L)$ is an infinite-dimensional quotient of F which is separable as the continuous image of the separable space $(F \oplus L)/M$. ■

7.4.11 Proposition *For an (LF)-space (E,τ), the following statements are equivalent:*

(i) (E,τ) has a defining sequence each of whose members has a separable quotient.

(ii) (E,τ) is not isomorphic to $F \oplus \Phi$, where F is any Fréchet space not having a separable quotient.

Proof. First, we consider the case where $E \cong F \oplus \Phi$ for some Fréchet space F. Let $((E_n, \tau_n))_n$ be any defining sequence for (E,τ). If $\Phi = \underrightarrow{\lim} G_n$ (G_n finite-dimensional), a defining sequence for $F \oplus \Phi$ is given by $(F \oplus G_n)_n$ (cf. 7.3.5).

Since $(E_n)_n$ and $(F \oplus G_n)_n$ are equivalent by 7.1.15, there exists $n_o \in \mathbf{N}$ such that the codimension of F in E_n is finite for $n \geq n_o$. Let L_n be an algebraic and hence topological complement of F in E_n for $n \geq n_o$. Then $E_n \cong F \oplus L_n$ for $n \geq n_o$.
Now assuming (i), the spaces E_n can be chosen so as to have a separable quotient each. By 7.4.10 F, too, has a separable quotient.
Conversely, (ii), together with $E \cong F \oplus \Phi$ yields that F has a separable quotient. By the reverse implication of 7.4.10, each E_n $(n \geq n_o)$ has a separable quotient. This completes the proof for the case $E \cong F \oplus \Phi$.
If E is not of this form for any Fréchet space then, of course, (ii) is true. To justify (i)$\Leftrightarrow$(ii) also in this case, we have to prove (*i*). By 7.1.24 there exists a defining sequence $((E_n, \tau_n))_n$ of (E, τ) such that $\tau_{n+1} \mid_{E_n} \neq \tau_n$ for each $n \in \mathbf{N}$.
Denote by F_n the closure of E_n in (E_{n+1}, τ_{n+1}) $(n = 1, 2, ...)$. Then $((F_n, \tau_{n+1} \mid_{F_n}))_n$ is a defining sequence for (E, τ) since each F_n is continuously included in F_{n+1}. Moreover, the inclusion is proper: Suppose that $F_{n-1} = F_n$ for some n. Then $\overline{E_n}^{\tau_{n+1}} = E_n$ and 7.1.8, applied to $id : (E_n, \tau_n) \to (E_n, \tau_{n+1} \mid_{E_n})$ yields $\tau_{n+1} \mid_{E_n} = \tau_n$, a contradiction. This also shows that $E_n = F_n (= \overline{E_n}^{\tau_{n+1}})$ is impossible, i.e. that E_n is a (dense) proper subspace of F_n.
Since $(E_n)_n$ and $(F_n)_n$ are equivalent, we have $(E, \tau) = \underrightarrow{\lim}(F_n, \tau_{n+1} \mid_{F_n})$ as (LF)-space. Finally, each $(F_n, \tau_{n+1} \mid_{F_n})$ has a separable quotient by 7.4.7, (v)$\Rightarrow$(i). ∎

7.4.12 Corollary *Every non-strict (LF)-space E possesses a defining sequence each of whose members has a separable quotient.*

Proof. E cannot be isomorphic to $F \oplus \Phi$ (F an arbitrary Fréchet space) since $F \oplus \Phi$ is a strict (LF)-space. ∎

7.4.13 Theorem *If (E, η) is a barrelled lcs such that (E, τ) is an (LF)-space for some $\tau \geq \eta$, then (E, η) has a separable quotient.*

Proof. Let $(E, \tau) = \underrightarrow{\lim}(E_n, \tau_n)$. We may assume without loss of generality that (E, η) is quasi-Baire. (If not, then (E, η) contains a complemented copy of Φ by 6.2.8 and Φ is infinite-dimensional and separable). Hence some E_k is dense in (E, η). The inclusion $(E_k, \tau_k) \hookrightarrow (E, \eta)$ is not nearly open: Otherwise, $\eta \mid_{E_k} = \tau_k$ by Ptàk's open mapping theorem (cf. [50], IV.8.3). But then E_k is closed and dense in (E, η), so that $E = E_k$, a contradiction.
Thus there exists an absolutely convex τ_k-neighborhood V of 0 in E_k such that its closure $\overline{V}$ in (E, η) is not an η-neighborhood of 0. $F := sp(\overline{V}) \supseteq E_k$ is dense in

(E,η) but not barrelled since $\overline{V}$ is not an $\eta\mid_F$-neighborhood of 0. It follows that the codimension of F in E is uncountable (see 1.5.3), so that we can choose $j \in \mathbf{N}$ such that $F \cap E_j$ is infinite-codimensional in E_j.

Let $\{U_n \mid n \in \mathbf{N}\}$ be a fundamental system of absolutely convex 0-neighborhoods in (E_j, τ_j) with $U_n + U_n \subseteq U_{n-1}$ for $n \geq 2$. Since $F \cap E_j$ is a proper subspace of E_j, we can choose $x_1 \in U_1 \setminus F$ and $f_1 \in (E,\eta)'$ with $f_1(x_1) = 1$ and $f_1 \in \overline{V}^{\circ}$ (Hahn-Banach). Set $V_1 = \overline{V}$.

By induction, we will define $(x_n)_n$ in E_j, $(f_n)_n$ in $(E,\eta)'$ and η-closed, absolutely convex sets

$$V_n := V_1 + \left\{ \sum_{i=1}^{n-1} a_i x_i \mid \; | a_i | \leq 1 \right\}$$

such that for $n \in \mathbf{N}$

(i) $x_n \in U_n$, $x_n \notin sp(V_n)$

(ii) $f_i(x_n) = 0$ for $i < n$, $f_n(x_n) = 1$, $f_n \in V_n^{\circ}$.

To perform the induction step we employ the same argument as in the proof of 7.4.6, so we omit the details.

There is, however, one subtle difference worth pointing out: Each x_n is chosen in $U_n (\subseteq E_j)$ (due to the fact that $sp(V_n) \cap E_j$ has infinite codimension in E_j) and U_n is a neighborhood of 0 with respect to τ_j. The linear forms f_n, though, have to be chosen as continuous with respect to the weaker topology η. This is possible since all V_n are (absolutely convex and) closed with respect to η.

Choose $y \in V_1$ and set

$$a_1 = -f_1(y) \; , \; a_n = -f_n \left(y + \sum_{i=1}^{n-1} a_i x_i \right)$$

for $n \geq 2$. Then $| a_1 | \leq 1$ and if $| a_i | \leq 1$ for $1 \leq i \leq k$, then $y + \sum_{i=1}^{k} a_i x_i \in V_{k+1}$, implying $| a_{k+1} | \leq 1$. Hence $a_i x_i \in U_i$ for each i, so that the Cauchy series $\sum_{i=1}^{\infty} a_i x_i$ converges to some z in (E_j, τ_j).

Moreover, since $\eta\mid_{E_j} \leq \tau\mid_{E_j} \leq \tau_j$, $\sum_{i=1}^{n} a_i x_i$ converges to z also in (E,η). For $n \in \mathbf{N}$, we have

$$f_n(y+z) = f_n \left(y + \sum_{i=1}^{n-1} a_i x_i \right) + a_n f_n(x_n) + \sum_{i=n+1}^{\infty} a_i f_n(x_i) = -a_n + a_n + 0 = 0.$$

Hence $y + z \in N := \bigcap_{n=1}^{\infty} f_n^{-1}(0)$.
$(y + z - \sum_{i=1}^{n} a_i x_i)_n$ is a sequence in $N + sp(\{x_n \mid n \in \mathbf{N}\})$ which is η-convergent to y. Thus

$$E = \overline{F} = \overline{sp(V_1)} \subseteq \overline{N + sp(\{x_n \mid n \in \mathbf{N}\})},$$

i.e. $N + sp(\{x_n \mid n \in \mathbf{N}\})$ is dense in (E, η).
Denoting by $\pi : E \to E/N$ the canonical projection, it follows that

$$\pi(N + sp(\{x_n \mid n \in \mathbf{N}\})) = sp(\{x_n + N \mid n \in \mathbf{N}\})$$

is dense in E/N, which is therefore separable. Finally,

$$dim(E/N) = codim\left(\bigcap_{n=1}^{\infty} f_n^{-1}(0)\right) = \infty,$$

since the f_n are linearly independent. ∎

7.4.14 Corollary *Every (LF)-space has a separable quotient.* ∎

7.4.15 Proposition *Let M be a closed, proper subspace of the (LF)-space $(E, \tau) = \varinjlim(E_n, \tau_n)$. Let $(U_k)_k$ be a 0-neighborhood base in some (E_n, τ_n) such that $\overline{M + U_k}$ is a τ-neighborhood of 0 for $k = 1, 2, ...$ Then $M + U_k$ is a τ-neighborhood of 0 for each $k \in \mathbf{N}$.*

Proof. Let $\pi : E \to E/M$ be the canonical projection and set

$$f : E_n/(M \cap E_n) \to E/M \ , \ f(x + (M \cap E_n)) = x + M.$$

Then f is continuous and

$$\pi(\overline{U_k + M}) \subseteq \overline{\pi(U_k)} = \overline{f(U_k + (M \cap E_n))}.$$

Thus f is nearly open because π is open. Since $E_n/(M \cap E_n)$ is a Fréchet space it follows that f is open (see [50], IV.8.3). Therefore $f(U_k + (M \cap E_n)) = \pi(U_k)$ is a neighborhood of 0, implying that $\pi^{-1}(\pi(U_k)) = U_k + M$ is a τ-neighborhood of 0 for every k. ∎

7.4.16 Proposition *Let V be a closed, absolutely convex subset of the (LF)-space $(E, \tau) = \varinjlim(E_n, \tau_n)$ such that $F = sp(V)$ is dense and proper in E. Suppose that for some $p \in \mathbf{N}$, $\overline{V + W}$ is a τ-neighborhood of 0 for each τ_p-neighborhood W of 0. Then E has a quotient which is a separable, infinite-dimensional Fréchet space.*

Proof. We have here a situation similar to the one in the course of the proof of 7.4.13: Since F is dense and proper, V is not a neighborhood of 0 in F. Thus F is not barrelled and therefore it is of uncountable codimension in E (1.5.3). Hence $F \cap E_j$ is of infinite codimension in E_j for some $j \geq p$.
Continuing to proceed as in the proof of 7.4.13 we obtain a sequence $(x_n)_n$ and a closed subspace N of (E, τ) such that E/N is infinite-dimensional and separable and $V \subseteq N + A$, where

$$A = \left\{ \sum_{i=1}^{\infty} a_i x_i \mid \ | a_i | \leq 1 \text{ for every } i \in \mathbf{N} \right\}.$$

We claim that A is bounded in (E_j, τ_j): It is enough to show that

$$B = \left\{ \sum_{i=1}^{N} a_i x_i \mid N \in \mathbf{N} \ , \ | a_i | \leq 1 \ \forall i \right\}$$

is bounded in (E_j, τ_j) (observe that $A \subseteq \overline{B}^{\tau_j}$).
Let $(U_n)_n$ be as in the proof of 7.4.13 and fix $n \in \mathbf{N}$. Then $\sum_{i=n+1}^{N} a_i x_i \in U_n$ for every $N \geq n+1$ and $| a_i | \leq 1$. Choose $\lambda > 0$ such that $\lambda^{-1} x_i \in U_n$ for $i = 1, ..., n$. Then $\sum_{i=1}^{n} a_i x_i \in n\lambda U_n$ (for arbitrary $| a_i | \leq 1$). Thus, with $\mu = \max(1, n\lambda)$ we have $\sum_{i=1}^{N} a_i x_i \in \mu(U_n + U_n)$, so that $B \subseteq \mu U_{n-1}$.
Next we are going to show $N + E_j = E$ by means of 7.4.15. (Observe that $N + E_j = \bigcup_{n=1}^{\infty} n(N + U)$ for any τ_j-neighborhood U of 0 in E_j.) Then an appeal to 7.3.11 will yield the result.
So let U be any absolutely convex τ_j-neighborhood of 0 in E_j and choose $\alpha > 0$ with $A \subseteq \alpha U$. Since $j \geq p$, the inclusion $E_p \hookrightarrow E_j$ is continuous. Thus $W := U \cap E_p$ is a τ_p-neighborhood of 0 in E_p and

$$\overline{V + W} \subseteq \overline{V + U} \subseteq \overline{N + (1 + \alpha)U}.$$

(Remember $V \subseteq N + A$.) It follows that $\overline{N + (1 + \alpha)U}$ and therefore $\overline{N + U}$ is a 0-neighborhood in (E, τ). Hence, by 7.4.15, $E = N + E_j$, so that E/N is Fréchet by 7.3.11. ■

7.4.17 Theorem *Every $(LF)_3$-space $(E, \tau) = \varinjlim(E_n, \tau_n)$ has a separable, infinite-dimensional quotient which is a Fréchet space.*

Proof. We may assume without loss of generality that E_1 is dense in E (since E is not of type 1). There exists an absolutely convex τ_1-neighborhood U_1 of 0 such that $\overline{U_1}^\tau$ is not a τ-neighborhood of 0 in E. (Otherwise the inclusion $(E_1, \tau_1) \hookrightarrow (E, \tau)$ would be a topological homomorphism by [50], IV.8.3. But then E_1 would be complete and dense in E, contradicting $E_1 \subset E$.) $sp(\overline{U_1}^\tau)$ is dense in E and proper since (E, τ) is barrelled.

Suppose that E has no quotient as described in the hypothesis. By 7.4.16 there exists an absolutely convex τ_2-neighborhood U_2 of 0 in E_2 such that $\overline{U_1 + U_2}^\tau$ is not a neighborhood of 0 in (E, τ).

Thus we can inductively define a sequence $(U_n)_n$ such that U_n is an absolutely convex τ_n-neighborhood of 0 in E_n and $A_n = \overline{U_1 + U_2 + ... + U_n}^\tau$ is not a neighborhood of 0 in (E, τ) for each $n = 1, 2, ...$

But then $(nA_n)_n$ is an increasing sequence of rare, absolutely convex sets whose union is E, contradicting the fact that E is Baire-like. ■

7.4.18 Remark If E is a separable, infinite-dimensional Fréchet space and E is an $(LF)_i$-space $(1 \leq i \leq 3)$, then $E \times F$ is an $(LF)_i$-space (7.3.7) having an infinite-dimensional, separable Fréchet-quotient.

On the other hand, Φ and l_p^- provide examples of $(LF)_1$- and $(LF)_2$-spaces, respectively, no quotient of which is an infinite-dimensional Fréchet space (cf. [36], p.636).

7.5 Metrizable and Normable (LF)-Spaces

In order to achieve a uniform formulation of the main result of this section, we first introduce some notations:

A *projection* on an lcs E is a continuous linear map $P : E \to E$ with $P \circ P = P$.

A sequence of projections is called *orthogonal*, if $P_j \circ P_k = 0$ for $j \neq k$.

We say that a Fréchet space E *splits* if there exist closed, infinite-dimensional subspaces M and N of E such that E is the algebraic (and therefore the topological, cf. [50], II.2.1, Cor.3) direct sum of M and N: $E = M \oplus N$.

E splits into infinitely many parts $(M_n)_n$, if there exist sequences $(M_n)_n$ and $(N_n)_n$ of closed, infinite-dimensional subspaces of E such that $E = M_1 \oplus N_1$, $N_1 = M_2 \oplus N_2$, $N_2 = M_3 \oplus N_3$,... .

7.5.1 Proposition *For a Fréchet space E, the following statements are equivalent:*

(i) E splits into infinitely many parts.

(ii) There exists an orthogonal sequence $(P_n)_n$ of projections on E with infinite-dimensional ranges.

Proof. (i)⇒(ii) Let $(M_n)_n$ and $(N_n)_n$ be as above. For each $n \in \mathbf{N}$ we have

$$E = M_1 \oplus M_2 \oplus ... \oplus M_n \oplus N_n.$$

Denote by P_n the projection of E onto M_n $(n = 1, 2, ...)$. Then $(P_n)_n$ is an orthogonal sequence.

(ii)⇒(i) Set $M_1 = P_1(E)$ and $N_1 = P_1^{-1}(0)$. Then $E = M_1 \oplus N_1$ and both M_1 and $N_1 \supseteq P_2(E)$ are infinite-dimensional.

Suppose that the closed, infinite-dimensional subspaces $M_1, ..., M_n$ and $N_1, ..., N_n$ have already been defined such that $M_k = P_k(E)$ and $N_k = \bigcap_{i=1}^k P_i^{-1}(0)$ for $1 \leq k \leq n$ and $E = M_1 \oplus M_2 \oplus ... \oplus M_n \oplus N_n$.

Now N_n is invariant under P_{n+1}: If $P_i(x) = 0$ for $1 \leq i \leq k$ then also $P_i \circ P_{n+1}(x) = 0$ for $1 \leq i \leq k$. Thus $P'_{n+1} := P_{n+1}|_{N_n}: N_n \to N_n$ is a projection on N_n and generates a direct sum decomposition $N_n = M_{n+1} \oplus N_{n+1}$. Here, $M_{n+1} = P'_{n+1}(N_n) = P_{n+1}(E)$ and

$$N_{n+1} = P'^{-1}_{n+1}(0) = P^{-1}_{n+1}(0) \cap N_n = \bigcap_{i=1}^{n+1} P_i^{-1}(0).$$

Thus we have $E = M_1 \oplus M_2 \oplus ... \oplus M_{n+1} \oplus N_{n+1}$. Finally, M_{n+1} and $N_{n+1} \supseteq P_{n+2}(E)$ are infinite-dimensional and closed. ∎

7.5.2 Theorem *If a Fréchet space (F, σ) splits into infinitely many parts, each of which has a (Hausdorff, infinite-dimensional) separable quotient, then F contains a dense subspace F_o such that $(F_o, \sigma|_{F_o})$ is a (metrizable) (LF)-space.*

(By 7.5.1, the hypothesis of 7.5.2 is satisfied if and only if there exists an orthogonal sequence $(P_n)_n$ of projections on F such that each $P_n(F)$ has a (Hausdorff, infinite-dimensional) separable quotient.)

Proof. For each $n \in \mathbf{N}$ there exists a dense, proper subspace G_n of $P_n(F)$ and a topology $\tau_n \geq \sigma|_{G_n}$ such that (G_n, τ_n) is a Fréchet space (see 7.4.7). For $n, k \in \mathbf{N}$, set $F_n = P_n^{-1}(G_n)$ and $E_k = \bigcap_{n=k}^{\infty} F_n$.

By 7.1.12 (i), (F_n, ξ_n) is a Fréchet space for each n, where ξ_n denotes the projective topology with respect to $P_n : F_n \to (G_n, \tau_n)$ and $F_n \hookrightarrow (F, \sigma)$. For each $k \in \mathbf{N}$ let σ_k be the projective topology on E_k with respect to the inclusions $E_k \hookrightarrow (F_n, \xi_n)$ $(n \geq k)$. 7.1.12 (ii) yields that (E_k, σ_k) is a Fréchet space.

E_k is continuously included in E_{k+1} ($k = 1, 2, ...$). Moreover, E_k is properly contained in E_{k+1}: Choose $x \in P_k(F)\backslash G_k$. Then $x \notin F_k$ (otherwise we would have $x = P_k(x) \in G_k$). Hence $x \notin E_k$. Now let $n > k$. Then $P_n(x) = P_n \circ P_k(x) = 0 \in F_n$, so that $x \in E_{k+1}$.

We assert that E_1 is dense in (F, σ). Let $(U_p)_p$ be a neighborhood base of 0 in (F, σ) such that each U_p is closed and absolutely convex and $U_p + U_p \subseteq U_{p-1}$ for $p \geq 2$. Fix $x \in E_1$, $k \in \mathbf{N}$ and choose $x_p \in G_p$ with $(x_p - P_p(x)) \in U_{k+p}$ for $p = 1, 2, \ldots$. The Cauchy series $\sum_{n=1}^{\infty}(x_n - P_n(x))$ converges to some $z \in U_k$ and we set $y := x + z$. Since $(P_n)_n$ is orthogonal, for each $j \in \mathbf{N}$ we have

$$P_j(y) = P_j(x) + x_j - P_j(x) = x_j \in G_j,$$

so that

$$y \in \bigcap_{j=1}^{\infty} P_j^{-1}(G_j) = E_1.$$

Consequently, $(x + U_k) \cap E_1 \neq \emptyset$ and E_1 is dense in F.

Equip $F_o = \bigcup_{k=1}^{\infty} E_k$ with the finest locally convex topology σ_o such that all the inclusions $(E_k, \sigma_k) \hookrightarrow F_o$ are continuous. Then (F_o, σ_o) is an (LF)-space densely and continuously included in (F, σ). It remains to show that $\sigma \mid_{F_o} \geq \sigma_o$.

Let V be a closed, absolutely convex neighborhood of 0 in (F_o, σ_o) and fix $k_o \in \mathbf{N}$. $V \cap E_{k_o}$ is a σ_{k_o}-neighborhood of 0 in E_{k_o} and we can choose $p_o \geq k_o$ and a neighborhood U_o of 0 in (F, σ) such that

$$E_{k_o} \cap U_o \cap \left(\bigcap_{n=k_o}^{p_o} P_n^{-1}(0) \right) \subseteq V$$

(observe that if '0' is replaced by a 0-neighborhood in (G_n, τ_n), the set on the left side of the above inequation is a 0-neighborhood in E_{k_o} with respect to σ_{k_o}).

For $k_1 > p_o$, $V \cap E_{k_1}$ is a σ_{k_1}-neighborhood of 0 and therefore there exists $p_1 \geq k_1$ and a neighborhood U_1 of 0 in $(F, \sigma)$with

$$E_{k_1} \cap U_1 \cap \left(\bigcap_{n=k_1}^{p_1} P_n^{-1}(0) \right) \subseteq V.$$

Set $P = (\sum_{n=k_o}^{p_o} P_n) \mid_{E_{k_o}}$. Then $P \circ P = P$ and for $k_o \leq n \leq p_o$ we have

$$P_n(E_{k_o}) = P_n \left(\bigcap_{i=k_o}^{\infty} P_i^{-1}(G_i) \right) \subseteq P_n(P_n^{-1}(G_n)) = G_n \subseteq E_{k_o},$$

so that $P(E_{k_o}) \subseteq E_{k_o}$. Hence P is a continuous linear projection on $(E_{k_o}, \sigma\mid_{E_{k_o}})$. Since $(P_n)_n$ is orthogonal, $P^{-1}(0) = (\bigcap_{n=k_o}^{p_o} P_n^{-1}(0)) \cap E_{k_o}$. It follows that $U_o \cap P^{-1}(0) \subseteq V$ and

$$U_1 \cap P(E_{k_o}) \subseteq E_{k_1} \cap U_1 \cap \left(\bigcap_{m=k_1}^{p_1} P_m^{-1}(0) \right) \subseteq V,$$

because $E_{k_o} \subseteq E_{k_1}$ and $k_1 > p_o$.
$P^{-1}(0)$ and $P(E_{k_o})$ are topological complements in $(E_{k_o}, \sigma\mid_{E_{k_o}})$, so that

$$W = \frac{1}{2}(U_o \cap P^{-1}(0)) + \frac{1}{2}(U_1 \cap P(E_{k_o}))$$

is a neighborhood of 0 in $(E_{k_o}, \sigma\mid_{E_{k_o}})$. Since $W \subseteq \frac{1}{2}V + \frac{1}{2}V \subseteq V$, $\sigma\mid_{E_{k_o}} = \sigma_o\mid_{E_{k_o}}$. Since k_o was arbitrary, it follows that $\sigma\mid_{E_k} = \sigma_o\mid_{E_k}$ for each $k \in \mathbf{N}$.
Let U be a closed, absolutely convex neighborhood of 0 in (F, σ) with $U \cap E_1 \subseteq V$. Then for each $p \in \mathbf{N}$, $U \cap E_p$ is the closure of $U \cap E_1$ in $(E_p, \sigma\mid_{E_p})$ since E_1 is dense in $(E_p, \sigma\mid_{E_p})$ (see 1.4.9).
Hence $U \cap E_p$ is the closure of $U \cap E_1$ in $(E_p, \sigma_o\mid_{E_p})$ and since V is σ_o-closed, it follows that $U \cap E_p \subseteq V$ for each $p \geq 1$. Finally,

$$U \cap F_o = U \cap \left(\bigcup_{p=1}^{\infty} E_p \right) \subseteq V,$$

so that V is a $\sigma\mid_{E_o}$-neighborhood of 0 and therefore $\sigma\mid_{F_o} = \sigma_o$. ∎

7.5.3 Corollary *Every Fréchet space (F, σ) with an unconditional basis contains a dense subspace F_o such that $(F_o, \sigma\mid_{F_o})$ is a (metrizable) (LF)-space.*

Proof. Let $\{x_n \mid n \in \mathbf{N}\}$ be an unconditional basis for F and $\{S_k \mid k \in \mathbf{N}\}$ a partition of $\mathbf{N}$ into infinite, disjoint sets. For $k \in \mathbf{N}$ define

$$P_k : F \to F \ , \ P_k \left(\sum_{n=1}^{\infty} a_n x_n \right) = \sum_{n \in S_k} a_n x_n.$$

Then $(P_n)_n$ is an orthogonal sequence of continuous linear projections on F with infinite-dimensional ranges.
Moreover, each $P_k(F)$ is clearly separable and therefore has a separable quotient by the subspace $\{0\}$. ∎

7.5.4 Corollary *Each of the following spaces contains a dense (LF)-subspace:*

$$\omega,\ (s),\ c_o,\ l^p\ (1 \leq p \leq \infty),\ C[0,1] \text{ and } L^p[0,1]\ (1 \leq p < \infty).$$

Proof. The proof is carried out by specifying orthogonal sequences of projections $(P_n)_n$ whose ranges admit separable quotients for those spaces, to whom the hypothesis of 7.5.3 doesn't apply.

l^∞: Since l^∞ has a separable quotient (see [43]), this follows analogous to the proof of 7.5.3 (observe that $P_n(l^\infty) \cong l^\infty$ for each $n \in \mathbf{N}$).

$C[0,1]$: Choose sequences $(a_n)_n$, $(b_n)_n$, $(c_n)_n$ and $(d_n)_n$ in $[0,1]$ such that $([a_n, b_n])_n$ is a sequence of disjoint subintervals of $[0,1]$ and $a_n < c_n < d_n < b_n$ for each $n \in \mathbf{N}$. Set $P_n : C[0,1] \to C[0,1]$,

$$P_n(f)(t) := \begin{cases} f(t) \text{ for } t \in [c_n, d_n] \\ 0 \text{ for } t \notin (a_n, b_n) \\ \text{linear on } [a_n, c_n] \text{ and } [d_n, b_n] \end{cases}$$

Then $\|P_n(f)\|_\infty \leq \|f\|_\infty$ and $P_n \circ P_m = \delta_{mn} P_n$, i.e. $(P_n)_n$ is an orthogonal sequence of projections on $C[0,1]$. Furthermore,

$$P_n(C[0,1]) \cong C([c_n, d_n]) \cong C[0,1]$$

is infinite-dimensional and separable for each $n \in \mathbf{N}$.

The same projections yield the result for $L^p[0,1]$ ($1 \leq p < \infty$; recall that $L^\infty[0,1]$ is not separable). ■

7.5.5 Corollary *Every non-Banach Fréchet space E has a dense (LF)-subspace.*

Proof. By 7.4.4, E has a quotient isomorphic to ω. ω has a dense (LF)-subspace by 7.5.4 (see also 7.2.18). Hence 7.3.15 completes the proof. ■

7.5.6 Remark 7.5.3-7.5.5, together with the remarks following 7.2.10 yield large classes of spaces each of whose members contains a dense subspace which is Baire-like but not (db) (all Fréchet spaces with an unconditional basis, all non-Banach Fréchet spaces, etc.)

7.5.7 Theorem *For a Banach space (E, τ), conditions (i)-(v) of 7.4.7 are equivalent to:*

(vi) (E,τ) densely and continuously includes a normable (LF)-space (E_o,τ_o).

Proof. (i)$\Rightarrow$(vi) Let M be a closed subspace of (E,τ) such that E/M is a separable, infinite-dimensional Banach-space. Set $E_1 = E/M$ and let $(y_n)_n$ be a linearly independent sequence such that $sp(\{y_n \mid n \in \mathbf{N}\})$ is dense in E_1. For $n \in \mathbf{N}$, set $S_n = sp(y_1, ..., y_n)$. Choose $x_1 \neq 0$ in S_1 and $f_1 \in E_1'$ with $f_1(x_1) = 1$.
Since $dim(S_2) = 2$ and $codim(f_1^{-1}(0)) = 1$, there exist $x_2 \neq 0$ in $f_1^{-1}(0) \cap S_2$ and $f_2 \in E_1'$ with $f_2(x_1) = 0$ and $f_2(x_2) = 1$. By induction, we obtain a biorthogonal sequence $(x_n, f_n)_n$ in $E_1 \times E_1'$ and since $sp(x_1, ..., x_n) = S_n$ for each n, it follows that $sp(\{x_n \mid n \in \mathbf{N}\})$ is dense in E_1.
Set $z_n = \frac{x_n}{\|x_n\|}$ for $n = 1, 2, ...$ and $D = \{z_n \mid n \in \mathbf{N}\}$. Consider the linear map

$$\psi : l_D^1 \to E_1 \ , \ (\lambda_n)_n \to \sum_{n=1}^{\infty} \lambda_n z_n.$$

ψ is continuous, because

$$\|\sum_{n=1}^{\infty} \lambda_n z_n\| \leq \sum_{n=1}^{\infty} | \lambda_n | = \|\lambda\|_1$$

and injective since $(x_n, f_n)_n$ is biorthogonal. Moreover, $\psi(l_D^1) \supseteq sp(\{z_n \mid n \in \mathbf{N}\})$ is dense in E_1.
Thus E/M densely and continuously includes a copy of l^1. Let $\pi : E \to E/M$ be the canonical projection. By 7.1.12 (i), $F := \pi^{-1}(l^1)$ can be equipped with a locally convex topology σ such that (F,σ) is a Banach space continuously included in (E,τ) and such that $\pi \mid_F : (F,\sigma) \to l^1$ is a continuous surjection. (The projective topology with respect to a finite number of normable locally convex topologies is normable).
F is dense in E since l^1 is dense in E/M and π is open: If W is an open subset of E then

$$\pi(W \cap F) = \pi(W \cap \pi^{-1}(l^1)) = \pi(W) \cap l^1 \neq \emptyset$$

because l^1 is dense in E/M. Consequently, $W \cap F \neq \emptyset$.
Now l^1 has a dense (LF)-subspace by 7.5.4 and the assertion follows from 7.3.15.
(vi)$\Rightarrow$(ii) resp. (v) Let $(E_o, \tau_o) = \underrightarrow{\lim}(E_n, \tau_n)$ be a normable (LF)-space, densely and continuously included in (E,τ).
Consider $F = \bigcup_{n=1}^{\infty} \overline{E_n}^{\tau}$. If $(\overline{E_n}^{\tau})_n$ contains a strictly increasing subsequence, (ii) of 7.4.7 is satisfied. Otherwise, $(\overline{E_n}^{\tau})_n$ is eventually constant. Then $F = \overline{E_p}^{\tau}$ for some

$p \in \mathbf{N}$.
In this case, (E_p, τ_p) is a Fréchet space, dense in F (hence in E) with respect to τ, continuously included in (E, τ) and proper ($E_p \subset E_{p+1} \subsetneq E$). Thus we have shown (v). ■

7.6 Completions

In general, (LF)-spaces are not stable under the formation of completions: The completion of an $(LF)_3$-space is a Fréchet space and therefore not (LF). Thus no $(LF)_3$-space is complete. Several classical Fréchet and Banach spaces are completions of (LF)-spaces by 7.5.4.
On the other hand, many $(LF)_1$-spaces are complete, as is demonstrated by the class of strict (LF)-spaces and we already know examples of complete $(LF)_2$-spaces (cf. 7.2.15).

7.6.1 Example *An $(LF)_2$-space whose completion is not an (LF)-space.*

Let $(E, \tau) = \underrightarrow{\lim}(E_n, \tau_n)$ be a complete $(LF)_2$-space such that E_n is dense in (E_{n+1}, τ_{n+1}) for $n = 1, 2, \ldots$ (e.g. $E = l_p^-$). Then E_1 is dense in (E, τ) by 7.2.14.
Set $(F_n, \sigma_n) = (E_n, \tau_n) \times (E_n, \tau_n) \times \ldots$ and $(F, \sigma) = \underrightarrow{\lim}(F_n, \sigma_n)$. Denoting by π the product topology on $E \times E \times \ldots$, we have $\sigma \geq \pi\, |_F$. Hence σ is separated and (F, σ) is an (LF)-space. Moreover, F is a dense, proper subspace of $(E^{\mathbf{N}}, \pi)$.
We claim that $\pi\, |_F = \sigma$. Since F_1 is dense in both $(F, \pi\, |_F)$ and in (F, σ) (by 7.2.14), it is enough to show that $\pi\, |_{F_1} = \sigma\, |_{F_1}$: Suppose that $\pi\, |_{F_1} = \sigma\, |_{F_1}$ has already been proved. Then 7.2.16, applied to the diagram

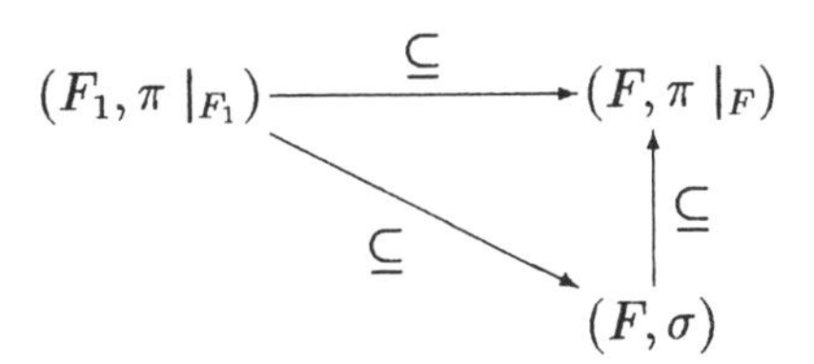

yields that $\sigma = \pi\, |_F$.
Obviously, $\sigma\, |_{F_1} \geq \pi\, |_{F_1}$. Conversely, let W be an absolutely convex neighborhood of 0 in (F, σ). Then

$$W \supseteq \Gamma_n(\underbrace{W_n \times \ldots \times W_n}_{p_n} \times E_n^{\mathbf{N}}),$$

where W_n is an absolutely convex neighborhood of 0 in (E_n, τ_n) and $(p_n)_n$ is a subsequence of $\mathbf{N}$. $V := \Gamma_n W_n$ is a 0-neighborhood in (E, τ), so that

$$U := \frac{1}{p_1+1}(\underbrace{V \times ... \times V}_{p_1} \times E^{\mathbf{N}}) \cap F$$

is a 0-neighborhood in $(F, \pi \mid_F)$. Let z be any member of $U \cap F_1$. Then

$$z = \frac{1}{p_1+1}(z_1, ..., z_{p_1}, x_{p_1+1}, x_{p_1+2}, ...),$$

where $z_1, ..., z_{p_1} \in E_1 \cap V$ and $x_{p_1+1}, x_{p_1+2}, ... \in E_1$. z_1 is of the form $\sum_{i=1}^N \lambda_i x_i$, where λ_i are scalars satisfying $\sum_{i=1}^N \mid \lambda_i \mid \leq 1$ and $x_i \in W_i$ (N depending on z, of course).

Since for each i,

$$(x_i, 0, 0, ...) \in \underbrace{W_i \times ... \times W_i}_{p_i} \times E_i^{\mathbf{N}},$$

we have

$$(z_1, 0, 0, ...) = \sum_{i=1}^N \lambda_i (x_i, 0, 0, ...) \in \Gamma_n(\underbrace{W_n \times ... \times W_n}_{p_n} \times E_n^{\mathbf{N}}) \subseteq W.$$

By a similar reasoning, we get that each $(0, ..., 0, z_k, 0, ...)$ belongs to W for $1 \leq k \leq p_1$. On the other hand,

$$(0, ..., 0, x_{p_1+1}, x_{p_1+2}, ...) \in \underbrace{W_1 \times ... \times W_1}_{p_1} \times E_1^{\mathbf{N}}.$$

Altogether, we have

$$z = \frac{1}{p_1+1}[(z_1, 0, 0, ...) + (0, z_2, 0, ...) + ... + (0, ..., 0, z_{p_1}, 0, 0, ...) +$$

$$+(0, ..., 0, x_{p_1+1}, x_{p_1+2}, ...)] \in \underbrace{\frac{1}{p_1+1}W + ... + \frac{1}{p_1+1}W}_{p_1+1} \subseteq W.$$

Therefore $U \cap F_1 \subseteq W$ and $\sigma \mid_{F_1} = \pi \mid_{F_1}$.

(F, σ) is not metrizable since (E, τ) isn't. Furthermore, since each F_n is dense in (F, σ), (F, σ) is an $(LF)_2$-space. The completion of (F, σ) is $(E^{\mathbf{N}}, \pi)$ which is not an (LF)-space by 7.3.9.

7.6.2 Example *An* $(LF)_1$*-space whose completion is not an (LF)-space.*

With (F, σ) as in 7.6.1, $F \times \Phi$ is an $(LF)_1$-space (7.3.6). Its completion $\widetilde{F} \times \Phi$ is not an (LF)-space since $(\widetilde{F} \times \Phi)/(\{0\} \times \Phi) \cong \widetilde{F}$ is neither an (LF)-space nor a Fréchet space (see 7.3.11).

7.6.1 and 7.6.2 are considerably narrowing the possibilities of proving general statements on the completion of (LF)-spaces.

In [20], J.M. GARCIA-LAFUENTE claimed that the completion of an (LB)-space is necessarily an (LB)-space or a Banach space. However, as professor S.A.SAXON pointed out to me, the proof given in [20] contains a gap (cf. the review of this article by J.BONET in MR 89e:46004). Thus the above assertion remains an open problem.

Moreover, this question is closely related to another classical problem, raised by A.GROTHENDIECK: Is every regular (LB)-space complete? (An (LB)-space $(E, \tau) = \underrightarrow{\lim}(E_n, \tau_n)$ is called regular if each bounded subset of (E, τ) is contained in some (E_n, τ_n) and bounded there.)

7.7 References

The theory of (LF)-spaces as presented in this chapter was decisively influenced by three articles by P.P.NARAYANASWAMI and S.A.SAXON ([35], [36] and [37]). Most of the results included here are taken from these papers.

To be more specific: 7.1.19 is taken from [36]. The special cases 7.1.12 (i) and (ii) of 7.1.11 were first published in [37]. 7.1.14 is essentially due to A.GROTHENDIECK ([22], p.16, cf. also [28], 19.5 (4)).

7.1.15-7.1.18 appeared in [37]. 7.1.21 and an earlier version of 7.1.24 were published without proof in [35] and [36].

7.2.1-7.2.8 and 7.2.14 are taken from [36], while 7.1.11-7.1.13 appeared in [36]. Example 7.2.17 is taken from [35]. (However, the words 'One easily sees that' in [35], p.69, Ex.1 appear to be slightly exaggerated in view of 7.2.16 and the proof of 7.2.17.)

7.3.1-7.3.10 go back to [36], the other results in section 3 appeared in [37].

7.4.1 and 7.4.2 due to M.EIDELHEIT can be found in [17], as was already mentioned. 7.4.6 was established by S.A.SAXON and A.WILANSKY in [49]. Concerning 7.4.7, cf. [49] and [37]. The equivalence of (iii) and (v) in 7.4.7 was first proved

by G.BENNET and N.J.KALTON ([4]). 7.4.11-7.4.18 are taken from [36].
All the results in section 5 appeared in [37].
7.6.1 was published without proof in [20]. Both examples 7.6.1 and 7.6.2 seem to be essentially due to S.A.SAXON.

References

[1] J.Arias de Reyna. *Dense hyperplanes of first category.*
Math.Ann.**249**, 111-114 (1980)

[2] J.Arias de Reyna. *Normed barely Baire spaces.*
Israel J.Math.**42**, 33-36 (1982)

[3] I.Amemiya,Y.Komura. *Über nicht-vollständige Montelräume.*
Math.Ann.**177**, 273-277 (1968)

[4] G.Bennet,N.J.Kalton. *Inclusion theorems for K-spaces.*
Can.J.Math.**25**, 511-524 (1973)

[5] J.Bonet,P.Pérez Carreras.*Barrelled locally convex spaces.*
North-Holland mathematical studies **131**, (1987)

[6] N.Bourbaki. *Espaces vectoriels topologiques.*
Hermann, Paris (1955)

[7] N.Bourbaki. *Intégration.*
Hermann, Paris (1956-1969)

[8] J.Cigler,V.Losert,P.Michor. *Banach modules and functors on categories of Banach spaces.*
Lecture notes in pure and applied Math.**46**, Marcel Decker, Inc (1979)

[9] M.De Wilde. *Sur les sous-espaces de codimension finie d'un espace linéaire à semi-normes.*
Bull.Soc.Roy.Sci.Liège **38**, 450-453 (1969)

[10] M.De Wilde,C.Houet. *On increasing sequences of absolutely convex sets in locally convex spaces.*
Math.Ann.**192**, 257-261 (1971)

[11] P.Dierolf,S.Dierolf,L.Drewnowski. *Remarks and examples concerning unordered Baire-like and ultrabarrelled spaces.*
Colloqu.Math.**39**, 109-116 (1978)

[12] J.Diestel,S.A.Morris,S.A.Saxon. *Varieties of locally convex topological vector spaces.*
Bull.Amer.Math.Soc.**77**, 799-803 (1971)

[13] J.Diestel,S.A.Morris,S.A.Saxon. *Varieties of linear topological spaces.*
Transactions Amer.Math.Soc.**172**, 207-230 (1972)

[14] J.Dieudonné. *Sur les propriétés de permanence de certains espaces vectoriels topologiques.*
Ann.Soc.Polon.Math.**25**, 50-55 (1952)

[15] N.Dunford,J.T.Schwartz. *Linear operators I: General Theory.*
Pure and Appl.Math.**7**, Interscience, New York (1958)

[16] R.E.Edwards. *Functional analysis.*
Holt,Rinehart and Winston, New York (1965)

[17] M.Eidelheit. *Zur Theorie der Systeme linearer Gleichungen.*
Studia Math.**6**, 139-148 (1936)

[18] K.Floret. *Folgenretraktive Sequenzen lokalkonvexer Räume.*
J.f.reine u. angewandte Math.**259**, 65-85 (1973)

[19] J.M.Garcia-Lafuente. *Countable-codimensional subspaces of c_o-barrelled spaces.*
Math.Nachr.**130**, 69-73 (1987)

[20] J.M.Garcia-Lafuente. *On the completion of (LF)-spaces.*
Monatshefte f.Math.**103**,115-120 (1987)

[21] A.Grothendieck. *Topological vector spaces.*
Gordon and Breach, New York-London-Paris(1973)

[22] A.Grothendieck. *Produits tensoriels topologiques et espaces nucléaires.*
Mem.Amer.Math.Soc.**16** (1955)

[23] J.Horvath. *Topological vector spaces and distributions.*
Addison-Wesley (1966)

[24] H.Jarchow. *Locally convex spaces.*
B.G.Teubner, Stuttgart (1981)

[25] T.Kato. *Perturbation theory for nullity, deficiency and other quantities of linear operators.*
J.d'Analyse Math.**6**, 273-322 (1958)

[26] V.Klee,A.Wilansky. *Research problems,#13.*
Bull.Amer.Math.Soc.**28**, 151 (1966)

[27] T.Komura,Y.Komura. *Über die Einbettung der nuklearen Räume in* $(s)^A$.
Math.Ann.**162**, 284-288 (1966)

[28] G.Köthe. *Topologische lineare Räume I.*
Springer Verlag, Berlin-Heidelberg-New York (1966)

[29] G.Köthe. *Topological vector spaces II.*
Springer Verlag, Berlin-Heidelberg-New York (1980)

[30] G.Köthe. *Die Bildräume abgeschlossener Operatoren.*
J.f.reine u.angew.Math.**232**, 110-111 (1968)

[31] E.Lacey. *Separable quotients of Banach spaces.*
Anais da Academia Brasileira de Ciências **44**, 185-189 (1972)

[32] M.Levin,S.A.Saxon. *Every countable-codimensional subspace of a barrelled space is barrelled.*
Proc.Amer.Math.Soc.**29**, 91-96 (1971)

[33] M.Levin,S.A.Saxon. *A note on the inhertitance properties of locally convex spaces by subspaces of countable codimension.*
Proc.Amer.Math.Soc.**29**, 97-102 (1971)

[34] J.Lindenstrauss,L.Tsafriri. *Classical Banach spaces*
Lecture Notes in Math.**338**, Springer Verlag, Berlin-Heidelberg-New York

[35] P.P.Narayanaswami, S.A.Saxon. *Metrizable (LF)-spaces, (db)-spaces and the separable quotient problem.*
Bull.Austral.Math.Soc.**23**, 65-80 (1981)

[36] P.P.Narayanaswami, S.A.Saxon. *(LF)-spaces, quasi-Baire spaces and the strongest locally convex topology.*
Math.Ann.**274**, 627-641 (1986)

[37] P.P.Narayanaswami,S.A.Saxon. *Metrizable (normable) (LF)-spaces and two classical problems in Fréchet (Banach) spaces.*
Studia Math.**93**, 1-16 (1989)

[38] J.C.Oxtoby. *Cartesian products of Baire spaces.*
Fundamenta Mathematicae **49**, 158-166 (1961)

[39] W.Robertson. *Completions of topological vector spaces.*
Proc.London Math.Soc.**8**, 242-257 (1958)

[40] D.J.Randtke. *On the embedding of Schwartz spaces into product spaces.*
Proc.Amer.Math.Soc.**55**, 87-92 (1976)

[41] W.J.Robertson,S.A.Saxon. *Dense barrelled subspaces of uncountable codimension.*
Proc.Amer.Math.Soc.**107**, 1021-1029 (1989)

[42] W.J.Robertson,I.Tweddle,F.E.Yeomans. *On the stability of barrelled topologies III.*
Bull.Austral.Math.Soc.**22**, 99-112 (1980)

[43] H.P.Rosenthal. *On quasi-complemented subspaces of Banach spaces.*
Journal of Functional Analysis **4**, 176-214 (1969)

[44] S.A.Saxon. *Embedding nuclear spaces in products of an arbitrary Banach space.*
Proc.Amer.Math.Soc.**34**, 138-140 (1972)

[45] S.A.Saxon. *Nuclear and product spaces, Baire-like spaces and the strongest locally convex topology.*
Math.Ann.**197**, 87-106 (1972)

[46] S.A.Saxon. *Some normed barrelled spaces which are not Baire.*
Math.Ann.**209**, 153-160 (1974)

[47] S.A.Saxon. *Two characterizations of linear Baire spaces.*
Proc.Amer.Math.Soc.**45**, 205-208 (1974)

[48] S.A.Saxon,A.R.Todd. *A property of locally convex Baire spaces.* Math.Ann.**206**, 23-34 (1973)

[49] S.A.Saxon,A.Wilansky. *The Equivalence of some Banach space problems.* Colloquium Math.**37**, 217-226 (1977)

[50] H.H.Schaefer. *Topological vector spaces* Springer Verlag, Berlin-Heidelberg-New York (1986)

[51] H.Schubert. *Topologie.* B.G.Teubner (1975)

[52] T.Terzioğlu. *On Schwartz spaces.* Math.Ann.**182**, 236-242 (1969)

[53] M.Valdivia. *Absolutely convex sets in barrelled spaces.* Ann.Inst.Fourier,Grenoble **21**,2, 3-13 (1971)

[54] M.Valdivia. *Some examples on quasi-barrelled spaces.* Ann.Inst.Fourier **22**, 21-26 (1972)

[55] M.Valdivia. *On suprabarrelled spaces.* Lecture Notes in Math.**843**, Springer, 572-580 (1981)

[56] M.Valdivia. *Topics in locally convex spaces.* North-Holland mathematical studies **67** (1982)

[57] A.Wilansky. *Functional Analysis.* Blaisdell, New York (1964)

[58] A.Wilansky. *Topics in functional analysis.* Lecture Notes in Math.**45**, Springer (1967)

Index